Who's Who
in Fluorescence 2007

Who's Who
in Fluorescence 2007

by

Chris D. Geddes[1] and Joseph R. Lakowicz[2]

[1]Founding Editor-in-Chief: Institute of Fluorescence,

[2]Co-Editor: Center for Fluorescence Spectroscopy,

University of Maryland, Baltimore, USA.

 Springer

Dr Chris D. Geddes
Institute of Fluorescence, N249,
Medical Biotechnology Center,
University of Maryland Biotechnology Institute,
725 West Lombard Street,
Baltimore, MD 21201, USA.
geddes@umbi.umd.edu

Dr Joseph R. Lakowicz
Centre for Fluorescence Spectroscopy,
University of Maryland School of Medicine,
725 West Lombard Street,
Baltimore, MD 21201, USA.
lakowicz@cfs.umbi.umd.edu

ISBN-10: 0-387-69796-9
ISBN-13: 978-0-387-69796-3
eISBN-10: 0-387-69798-5
eISBN-13: 978-0-387-69798-7

Printed on acid-free paper.

9 8 7 6 5 4 3 2 1

springer.com

To all those who employ fluorescence in their working lives,

We hope you find this volume a useful resource,

Chris D. Geddes,
Baltimore, USA, September 2006.

Preface

Celebrating 5 years of the WWIF Volume

The Who's Who in Fluorescence 2007 is the 5th volume of the Who's who series. The previous four volumes (2003 - 2006) have been very well received indeed, with 1000's of copies being distributed around the world, through conferences and workshops, as well as through internet book sites.

In the last 5 years a great many of you have sent comments and suggestions, we thank you all. We continue to accommodate as many of these as possible into the new volumes. This new 2007 volume features some 405 entries from no fewer than 35 countries worldwide, as compared to 366 entries in the 2006 volume. We have received 45 new entries this year, and deleted 6 entries that were not updated by contributors from past years, some 62 entries being deleted in the 2006 volume. Some older entries were again retained by the editors.

In addition, we have a continued strong company support, which will enable us to further disseminate the volume in 2007-2008. In this regard we especially thank the instrumentation companies for their continued support, where without their financial contributions, it is likely that the volume would not be the success it is today.

We introduced a new author publication statistic into the 2005 volume, the Author Impact Measure (AIM) number. While voluntary, this number is intended to reflect an author's publication progress in past years. The AIM number simply summates the *impact number* (from the ISI database) of Journals published in, in that year, multiplied by the frequency of those publications. From those who chose to participate, we can see most impressive AIM numbers. We have subsequently additionally listed these at the front of the volume for simple comparison between contributors. Last year, some 88 AIM numbers were submitted and listed, 37 the year before. This year, the number submitted has risen again to 106 entries, greater than 25 % of all contributors.

Finally, we would like to thank Caroleann Aitken, The *Who's Who in Fluorescence Co-ordinator* for both the architecture and the typesetting of the entire volume in a timely fashion. Thanks also go to Aaron Johnson at Springer, for helping to make this volume possible, many thanks.

Dr Chris D. Geddes,
Professor,
Baltimore, USA.
November 2006.

Who's Who
in Fluorescence 2007

Contents

Personal Entries

Company & Institution Entries

AIM (Author Impact Measure) Number

M. P. Aguilar-Caballos,	AIM 2004 = 11.5	1
Wajih Al-Soufi,	AIM 2005 = 22.6	4
Ricardo F. Aroca,	AIM 2005 = 45.6	7
Kadir Aslan,	AIM 2005 = 55.1	8
Gary A. Baker,	AIM 2003 = 28.4	10
Susan L. Bane,	AIM 2004 = 14.7	11
Denis Boudreau,	AIM 2006 = 16.8	19
Frank V. Bright,	AIM 2004 = 35.0	19
Haishi Cao,	AIM 2006 = 4.4	25
Zorgan G. Cerovic,	AIM 2004 = 4.2	26
Amitabha Chattopadhyay,	AIM 2004 = 61.1	29
Mustafa H. Chowdhury,	AIM 2006 = 13.1	31
Suresh Das,	AIM 2004 = 18.0	34
Amilra P. de Silva,	AIM 2004 = 23.5	36
Alexander P. Demchenko,	AIM 2003 = 26.9	38
Wen-Ji Dong,	AIM 2005 = 9.1	41
Kim Doré,	AIM 2006 = 8.2	42
Guy Duportail,	AIM 2004 = 12.7	44
Richard H. Ebright,	AIM 2005 = 86.0	47
Yves Engelborghs,	AIM 2004 = 55.5	49
Sergie A. Eremin,	AIM 2005 = 11.8	50
Jose Paulo S. Farinha,	AIM 2003 = 12	52
Maria L. Ferrer,	AIM 2004 = 9.9	54
Ehud Gazit,	AIM 2005 = 65.6	59
Chris D. Geddes,	AIM 2005 = 102.0	60
Marcelo H. Gehlen,	AIM 2006 = 16.8	60
Augustina Gomez-Hens,	AIM 2004 = 14.1	62
Karl Otto Greulich,	AIM 2005 = 15.2	63
Oleksiy V. Grygorovych,	AIM 2004 = 1.0	65
Eugene E. Gussakovsky,	AIM 2005 = 14.3	59
Carlos Gutiérrez-Merino,	AIM 2005 = 12.6	66
Harri O. Hakala,	AIM 2004 = 7.7	67
Gregory S. Harms,	AIM 2005 = 16.3	69
Stefan W. Hell,	AIM 2005 = 41.8	71
Albin Hermetter,	AIM 2004 = 33.6	73
Jordi Hernandez-Borrell,	AIM 2005 = 17.9	74
Graham Hungerford,	AIM 2005 = 9.9	78
Tony D. James,	AIM 2005 = 20.7	80
Arthur E. Johnson,	AIM 2004 = 85.0	82
Carey K. Johnson,	AIM 2005 = 22.9	82
Pedro A. S. Jorge,	AIM 2006 = 2.7	84
Nicoletta Kahya,	AIM 2005 = 39.0	84

MAF 10

The 10th Conference on Methods an Applications of Fluorescence: Spectroscopy, Imaging and Probes (MAF 10), Salzburg, Austria, 9 – 12 Sep. 2007.

The biannual MAF conferences, held in Graz (1989 and 1991), Prague (1993), Cambridge (1995), Berlin (1997), Paris (1999), Amsterdam (2001), Prague (2003), and Lisbon (2005), have acquired high reputation. They gather researchers dealing with all aspects of fluorescence. The MAF 10 Conference will cover new techniques and methods, and their applications to a wide range of fields.

The MAF 10 Conference is of interest to all of those that work in fluorescence in some way, and will cover physical, chemical, biological, drug discovery, pharmacological, diagnostic, medical imaging, and material science aspects. It will include contributions on new spectroscopic methods and techniques, the development and application of fluorescent probes, and new techniques and applications of fluorescence imaging.

Thirty speakers of highest standing have been invited that will present their latest results. Six communications will be selected from the Abstracts submitted, and the respective authors will be invited to give a talk. Please submit your Abstract as soon as possible.

MAF 10 is expected to attract a large number of participants, owing not only to the cutting-edge scientific content, but also to the attractive location and social programmed (including a classical music concert). The number of participants in the conference is limited to 350, and early registration is strongly recommended for this reason.

Contact: http://maf10.maf-sip.net/

MAF 10

Date submitted: 31st October 2006

A. Ulises Acuña, Ph.D.

Department of Biophysics,
Instituto de Química-Física "Rocasolano" C.S.I.C.,
119 Serrano, Madrid-28006,
Spain.
Tel: +34 91 561 9400   Fax: +34 91 564 2431
roculises@iqfr.csic.es
www.iqfr.csic.es

Specialty Keywords: Polarised luminescence spectroscopy fluorescent bioprobes, membrane structure and dynamics.

Research: Synthesis of new fluorescent labels, probes and analogs of anticancer   and antiparasite drugs. Theory of rotational depolarisation of luminescence. Probing protein and lipid membrane dynamics with time-resolved fluorescence, phosphorescence and T-T dichroism. Theory of T-T energy transfer and excited-state proton transfer. The history of solution fluorescence.

-Fluorescent phenylpolyene analogs of the ether phospholipid edelfosine for the selective labeling of cancer cells. Quesada, E., Delgado, J, Gajate, C. Mollinedo, F. Acuña, A.U. and Amat-Guerri, F. (2004) *J. Med. Chem.* 47, 5333;

-Dynamics of bolaamphiphilic fluorescent polyene in lipid bilayers from polarization emission spectroscopy. Acuña, A.U., Amat-Guerri, F. Quesada, E.Vélez, M. (2006) *Biophys. Chem.* 122, 27.

Date submitted: 1st July 2005

M. P. Aguilar-Caballos, Ph.D.

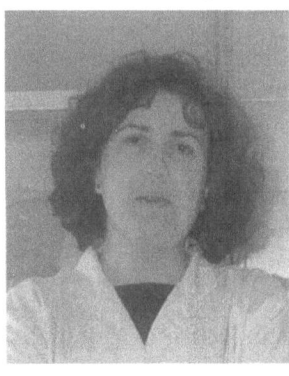

Department of Analytical Chemistry,
Faculty of Sciences, Campus of Rabanales, Anexo C-3,
University of Córdoba,
140741-Córdoba, Spain.
Tel: 3495 721 8645   Fax: 3495 721 8644
qa2agcam@uco.es

Specialty Keywords: Lanthanide, Fluoroimmunoassay, NIR dyes.
AIM 2004 = 11.5

Some topics of interest are the use of lanthanide chemistry, such as dry-reagent chemical technology or NIR emitting lanthanide ions among others, and also the study of new reactions to increase the reactivity of long-wavelength fluorophores. Research is also focused on the development of new homogeneous fluoroimmunoassay methods using kinetic methodology and immunoaffinity chromatography methods with fluorescence detection. These have been applied to different areas such as clinical, environmental and food analysis.

M.P. Aguilar-Caballos and A. Gómez-Hens (2004). Long-wavelength flurophores: new trends in their analytical use. *Trends Anal. Chem.* 23(2), 127-136.

Date submitted: 17th April 2006

Ramesh C. Ahuja, Ph.D.

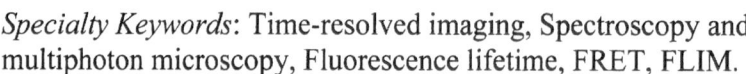

time-resolved imaging & microscopy

TauTec LLC.,
9140 Guilford Road, Suite O,
Columbia, MD,
USA.
Tel: 301 725 7441   Fax: 301 725 2941
rahuja@tautec.com
www.tautec.com

Specialty Keywords: Time-resolved imaging, Spectroscopy and multiphoton microscopy, Fluorescence lifetime, FRET, FLIM.

TauTec offers state-of-the-art, ultrahigh repetition rate (up to 110MHz), picosecond gated (down to 50ps), gain modulated (up to 1GHz) PicoStar ICCD cameras, ultrafast readout EMCCD cameras, modular multifocal multiphoton TriMScope workstations for real-time 3D fluorescence microscopy with time-lapse, ratio imaging, 2D and 3D kinetics, FLIM, FRET, FRAP, anisotropy and spectral imaging functionalities, TauScope for wide-field time or frequency domain FLIM/FRET based on pulsed Laser/LED excitation, time-gated Raman imaging and spectroscopy systems, plasma kinetics spectroscopy, gating and ranging LIDAR.

Date submitted: Editor Retained.

Engin U. Akkaya, Ph.D.

Department of Chemistry,
Middle East Technical University,
06531 Ankara,
Turkey.
Tel: 90 312 210 5126   Fax: 90 312 210 1280
akkayaeu@metu.edu.tr

Specialty Keywords: Fluorescent chemosensors, Molecular logic gates, Molecular devices.

Current interests: Design and synthesis of novel fluorogenic and chromogenic chemosensors for cations, anions and carbohydrates. Novel sensing schemes. Calixarene-based ion-pair sensors and allosterical modulation of binding interactions. Oxidative PET and cation/anion modulation of oxidative PET. Antenna systems. Diazapyrenium-based fluorescent pseudorotaxanes. Novel and efficient sensitizers for photodynamic therapy. Fluorescent chemosensors for dopamine.

-C. N. Baki and Engin U. Akkaya (2001). Boradiazaindacene appended calix[4]arene:
Fluorescence sensing of pH near neutrality, *J. Org. Chem.* 66, 1512-1513.
-B. Turfan and Engin U. Akkaya (2002). Modulation of Boradiazaindacene Emission by Cation Mediated Oxidative PET, *Organic Lett.* 4, 2857-2859.

Date submitted: 27th July 2004   **Jihad-René Albani, Ph.D.**

Laboratoire de Biophysique Moléculaire,
Université des Sciences et Technologies de Lille,
Bâtiment C6,
59655 Villeneuve d'Ascq Cédex, France.
Tel: 33 32 033 7770
Jihad-Rene.Albani@univ-lille1.fr

Specialty Keywords: Structure, Dynamics, Fluorescence fingerprints.

We characterize structure and dynamics of proteins. For example, we were able to characterize the global spatial structure of α_1 – acid glycoprotein showing the presence of a pocket where ligands can bind (Albani, 2004, Carbohydrate Research). Also, we showed that the carbohydrate residues of the protein possess a spatial structure (Albani et al. 2000, Carbohydrate Research). Also, we apply fluorescence to characterize species and varieties in animals and vegetals (Albani et al. 2003, Photochem. Photobiol).

-J. R. Albani. 2004. Structure and Dynamics of Macromolecules: Absorption and Fluorescence Studies. Book in English (418 Pages) published by Elsevier Sciences Ltd.
-J. R. Albani. 2001. Absorption et Fluorescence: Principes et Applications. Book in French (248 pages) published by Lavoisier-Tec et Doc.

\Date submitted: 11th August 2005   **Annette Alfsen, M.D., Ph.D.**

Cell Biology, Institut Cochin,
Université René Descartes,
22, rue Méchain, 75014 - Paris,
France.
Tel: 33 14 051 6445   Fax: 33 14 051 6454
alfsen@cochin.inserm.fr

Specialty Keywords: Cell biology-virus entry, Membranes, Biophysics.

Our domain of research is dealing with the biology of the epithelial cell and its interaction with the virus HIV budding from PBMC infected cell. The formation of a viral synapse between both cells and the different intermediate molecules acting as receptors of the viral envelope has been the topic of our last paper.

-Nature,Molecular cell biol. Review. Bomsel M. and Alfsen A. 2003,vol.4 nb.1
-HIV-1-infected blood mononuclear cells form an integrin- and agrin-dependent viral synapse to induce efficient HIV-1 transcytosis across epithelial cell monolayer. 2005 Annette Alfsen [1], Huifeng Yu [1], Aude Magérus-Chatinet [1], Alain Schmitt [#], and Morgane Bomsel . Mol.Cell.Biol. [1,]

Date submitted: 8th June 2006

Ron R. Allison, M.D.

Radiation Oncology,
ECU School of Medicine,
600 Moye Blvd, Greenville,
Pitt County, 27858-4345, USA.
Tel: 252 744 2900   Fax: 252 744 2812
Allisonr@mail.ecu.edu
www.ecu.edu/radonc

Specialty Keywords: Cancer, Photodynamic therapy, Optical biopsy.

Research interests are photodynamic therapy optimization for oncology patients, refine and improve therapy both by clinical modifications and dosimetry enhancement. We have the largest number of chest wall recurrence patients treated with PDT.

-Allison RR, Mota HC, and Sibata CH: Clinical History of PDT in North America: An Historical Overview. Photodiagnosis and Photodynamic Therapy, 1, 2005:263-277.

-Allison R, Sibata C, Mang TS, Bagnato VS, Downie GH, Hu XH, and Cuenca R: Photodynamic Therapy for Chest Wall Recurrence from Breast Cancer. Photodiagnosis and Photodynamic Therapy, 2004, 1,157-171.

Date submitted: 1st June 2005

Wajih Al-Soufi, Ph.D.

Universidad de Santiago de Compostela,
Facultad de Ciencias, Departamento de Química Física,
Campus Universitario s/n, E-27002 Lugo,
Spain.
Tel: +34 98 228 5900 Fax: +34 98 228 5872
alsoufi@lugo.usc.es

Specialty Keywords: Fluorescence, Data analysis, FCS.
AIM 2005 = 22.6

Study of the influence of confined media on proton transfer and charge transfer processes. Supramolecular Dynamics. Development and implementation of new data analysis methods for steady state and time resolved fluorescence data.

-W. Al-Soufi, B. Reija, M. Novo, S. Felekyan, R. Kühnemuth, and C.A.M. Seidel (2005) Fluorescence Correlation Spectroscopy, a Tool to investigate Supramolecular Dynamics: Inclusion Complexes of Pyronines with Cyclodextrin. *JACS,* 127, 8775-8784.
-B. Reija, W. Al-Soufi, M. Novo, J. Vázquez Tato (2005) Specific interactions in the inclusion complexes of Pyronines Y and B with β-cyclodextrin, *J. Phys. Chem. B*, 109, 1364-1370.

Date submitted: 15th June 2006

Marcel Ameloot, Ph.D.

University of Hasselt / Biomedical Research Institute,
Agoralaan, Building D,
B-3590 Diepenbeek,
Belgium.
Tel: 321 126 8546   Fax: 321 126 8599
marcel.ameloot@uhasselt.be
lucinfsr.uhasselt.be:8090/wie_zijn_wij/MarcelAmeloot_CV.asp

Specialty Keywords: Microfluorimetry, Data analysis, Cell physiology.

The current focus is on microfluorimetric determination of membrane microdomains in living cells, the use of fluorescence in the development of electronic biosensors and data analysis of multidimensional fluorescence data surfaces.

-Cooreman P, Thoelen R, Manca J, vandeVen M, Vermeeren V, Michiels L, Ameloot M, Wagner P. (2005) Biosens Bioelectron. 20 2151-6.
Boens N, Novikov E, Szubiakowski J, Ameloot M (2005) J Phys Chem A. 109(51) 11655-11664.
-Maria A. Acasandrei, Robert E. Dale, Martin vandeVen, Marcel Ameloot (2006) Chem. Phys. Lett. 419 469-473.

Date submitted: 8th June 2005

John E. Anderson, Ph.D.

U.S. Army Engineering Research and Development Center,
Fluorescence Remote Sensing Lab,
USAERDC-TEC, 7701 Telegraph Road, Alexandria,
Virginia, 22315 USA.
Tel: 703 428 8203   Fax: 703 428 8176
John.anderson@erdc.usace.army.mil
www.tec.army.mil

Specialty Keywords: Fluorescence Sensing, Molecular Imprinted Polymers, laser-induced fluorescence.

The Fluorescence Spectroscopy Lab at ERDC is engaged in basic and applied research in fluorescence sensing. This research is focused on the development and modeling of fluorophores for recovery by remote sensing. The Lab guides the development of organic (living) and inorganic materials that may be used for the targeting and detection of harmful agents or environmental threats of relevance to the military or civil communities. The Lab has the capabilities to measure both steady-state and lifetime (decay) fluorescence spectra for fluorophores using state-of-the art spectrometers and lasers including a frequency domain lifetime spectrofluorometers. The Lab has recently incorporated instruments to characterize near infrared fluorescence and apply mulit-photon excitation to targets of interest.

Date submitted: 26th April 2006

David L. Andrews, Ph.D.

Nanostructures and Photomolecular Systems,
School of Chemical Sciences,
University of East Anglia,
Norwich, NR4 7TJ, U.K.
Tel: +44 160 359 2014   Fax: +44 160 359 2003
david.andrews@physics.org
www.uea.ac.uk/~c051

Specialty Keywords: Quantum Electrodynamics, Resonance Energy Transfer, Multiphoton Absorption, Optical Forces.
Research in Andrews's group centers on condensed phase and molecular photophysics, based on the unified theory of resonance energy transfer. This group was first to identify and predict characteristics of two-photon FRET, anticipating experiments on biological systems. In ongoing projects the group works on directed energy transfer and energy harvesting in optically nonlinear nanostructures, optical switching, and laser-induced inter-particle forces.

-D. L. Andrews and D.S. Bradshaw, Optically nonlinear energy transfer in light-harvesting dendrimers, *J. Chem. Phys.* 121, 2445-2454 (2004).
-D. L. Andrews and R.G. Crisp, Theory of directed electronic energy transfer, *J. Fluor.* 16, 1-9 (2006).

Date submitted: Editor Retained.

Pavel Anzenbacher, Ph.D., D.Sc.

Inst. of Pharmacology, Faculty of Medicine,
Palacky University,
Hnevotinska st. 3, Olomouc,
CZ-779 00, Czech Republic.
Tel: +420 58 563 2569   Fax: +420 58 563 2966
anzen@tunw.upol.cz

Specialty Keywords: Protein conformation, Tryptophans, Heme enzymes.
Active sites of cytochromes P450 and other heme enzymes differ in amino acid residues to reflect their function and specificity. Tryptophan fluorescence is studied by stationary approach as well as by time-resolved techniques. Interaction with enzyme substrates often produce fluorescence changes which are characteristic for different cytochrome P450 enzymes. FCS gives then information on changes in protein aggregation and overall conformation.
-R. Lange, Anzenbacher P., Müller S., Maurin L., Balny C. (1994) Interact. of tryptophan residues in cytochrome P450scc with a fluorescence quencher *Eur.J.Biochem.* 226, 963-970.
-Bemeš M., Hudeček J., Anzenbacher P., Anzenbacher P., Hof M. (2001) Coumarin 6, resorufins and flavins: Suitable chromophores for FCS of biol. molecules.*Coll.Czech.Chem.Commun.*66, 855-869.

Date submitted: 26th July 2004

Jutta Arden-Jacob, Ph.D.

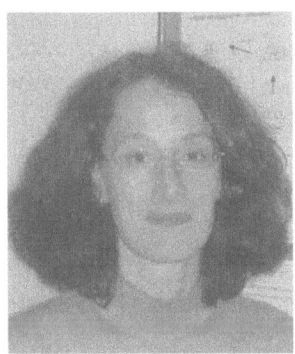

ATTO-TEC GmbH.,
Am Eichenhang 50,
D-57076 Siegen,
Germany.
Tel: +49 (0) 271 740 4022
arden-jacob@atto-tec.de
www.atto-tec.com

Specialty Keywords: Fluorescent dyes, Biolabelling, Red-absorbing Chromophors.

My research is focused on the chemical synthesis and characterization of new red-absorbing fluorophors. I am particularly interested in new fluorescent dyes which are suitable for biolabelling.

-J. Arden-Jacob, J. Frantzeskos, N.U. Kemnitzer, A. Zilles, and K.H. Drexhage (2001). New fluorescent markers for the red region *Spectrochim. Acta A* 57(11), 2271-2283.

-J. Arden-Jacob, N.J. Marx, and K.H. Drexhage (1997). New fluorescent probes for the red spectral region *J. Fluoresc.* 7(1), 91S-93S.

Date submitted: 20th September 2005

Ricardo F. Aroca, Ph.D.

Materials Surface Science Group,
Faculty of Science,
University of Windsor, Windsor,
Ont., N9B 3P4, Canada.
Raroca1@cogeco.ca

Specialty Keywords: Surface-Enhanced Spectroscopy,
Langmuir-Blodgett Films.
AIM 2005 = 45.6

Nanostructures for surface-enhanced spectroscopy. Surface enhanced-vibrational spectroscopy, surface enhanced fluorescence and nanostructure fabrication with appropriate optical properties to produce enhanced optical fields.

-Alvarez-Puebla, R.A., E. Arceo, P.J.G. Goulet, J.J. Garrido, and R.F. Aroca, *Role of Nanoparticle Surface Charge in Surface-Enhanced Raman Scattering.* Journal of Physical Chemistry B, 2005. 109(9): p. 3787-3792.

-Aroca, R.F., D. Ross, and C. Domingo, *Surface enhanced infrared spectroscopy (Review paper and Cover Art).* Applied Spectroscopy, 2004. 58: p. 324-338A.

Date submitted: 3rd August 2006

Kadir Aslan, Ph.D.

Institute of Fluorescence,
Laboratory for Advanced Medical Plasmonics (LAMP),
Medical Biotechnology Center, N249,
University of Maryland Biotechnology Institute,
725 West Lombard St., Baltimore, Maryland 21201, USA.
Tel: 410 706 4566   Fax: 410 706 4600
aslan@umbi.umd.edu

Specialty Keywords: Nanotechnology, Biosensors, Surface Chemistry.
AIM 2005 = 55.1

My research focuses on the development and application of plasmonic / fluorescence-based biosensors using noble metallic nanoparticles. I am also interested in self-organization of the nanoparticles on surfaces using specific biological interactions as well as other aspects of the subject nanotechnology.

-Aslan, K.; Huang, J.; Wilson, G.W.; Geddes, C.D. "*Metal-Enhanced Fluorescence-Based RNA Sensing*", J. Am. Chem. Soc. (2006), 128(13), 4206-4207.
-Aslan, K.; Geddes, C.D. "Microwave-Accelerated Metal-Enhanced Fluorescence: A New Platform Technology for Ultra-Fast and Ultra-Bright Assays", Anal. Chem. (2005), 77(24), 8057-8067.

Date submitted: 5th September 2006

Salvatore H. Atzeni, Ph.D.

Optical Spectroscopy Division,
HORIBA Jobin Yvon,
3880 Park Avenue,
Edison, NJ 08820-3012 USA.
Tel: 732 494 8660   Ext: 131   Fax: 732 549 5125
sal.atzeni@jobinyvon.com
www.jobinyvon.com

Specialty Keywords: Imaging CCD, Anisotropy, TCSPC.

Graduate research focused on *Anisotropic Rotations of Perylene in Anisotropic Media.*

Dr. Atzeni is the Director of the Optical Spectroscopy Division of HORIBA Jobin Yvon. The Optical Spectroscopy Division specializes in plug and play components for spectroscopy. We offer light sources, small monochromators, high-resolution spectrometers, spectographs, imaging spectrographs, single-channel detectors, multichannel detectors (CCD, iCCD, IGA) and optical accessories all designed to work together as a complete system for fluorescence, Raman, and light-measurement applications.

Date submitted: Editor Retained.

Luis A. Bagatolli, Ph.D.

Memphys - Center for Biomembrane Physics,
Department of Physics, University of Southern Denmark,
Campusvej 55, DK-5230 Odense M,
Denmark.
Tel: +45 65 50 3476   Fax: +45 66 15 8760
bagatolli@memphys.sdu.dk
www.memphys.sdu.dk/

Specialty Keywords: Multiphoton Microscopy, Polarity
Sensitive Probes, Lipid, Lipid and Lipid, Protein Interactions.

My primary research goal is to study lipid / lipid and lipid protein interactions in natural and model membranes. The fluorescence parameters measured in traditional experiments involving liposome solutions can be measured at the level of single vesicles using fluorescence microscopy. Using this last approach is possible to establish a correlation between the microscopic organization on the surface of single vesicles with the physical parameters determined at molecular level on the lipid bilayer (lipid mobility, lipid hydration, etc).

-Bagatolli L.A. and E. Gratton. (2001) *J. of Fluorescence* 11:141-160.
-Sanchez S., L. A. Bagatolli, E. Gratton, T. Hazlett (2002) *Biophys. J.* 82:2232-2243.

Date submitted: 15th August 2003

Željko Bajzer, Ph.D.

Department of Molecular Biology,
and Biochemistry, Mayo Clinic Rochester,
200 First Street SW, Rm. 1611B GU,
Rochester, MN 55905, USA.
Tel: 507 284 8584   Fax: 507 284 9420
bajzer@mayo.edu

Specialty Keywords: Multiexponental Models, Parameter
Estimation, Deconvolution Methods.

My focus in the field of biological fluorescence is on investigation and development of methods for data analysis and on study of multiexponential models. Previous work: The Pade-Laplace method for the analysis of time and frequencey domain lifetime measurements; a model for tryptophan fluorescence decay in proteins; new methods for discretization of convolution integrals, yielding more accurate determination of lifetimes and anisotropy decay parameters. Recently: Application of stretched exponential models and fractal kinetics.

Date submitted: 6th June 2004

Gary A. Baker, Ph.D.

Oak Ridge National Laboratory,
P.O. Box 2008,
Oak Ridge, TN 37831,
USA.
Tel: 865 241 9361
bakerga@ornl.gov

Specialty Keywords: Nanomaterials, Ionic liquids, Biosensing.
AIM 2003 = 28.4

Current efforts in my research group focus on the following research topics: biospectroscopy in ionic liquids & confining media, nanoparticles for fluorescence sensing, photoactive dendrimers, and green solvent systems, including water-in-CO_2 microemulsions and ionic fluid media.

-S. Pandey et al., (2004). Generation and pH dependent superquenching of poly(amido) carboxylate dendrons hosting a single "focal point" pyrene. *Chem. Commun.,* 1318–1319.

-S. N. Baker, T. M. McCleskey, S. Pandey, and G. A. Baker, (2004). Fluorescence Studies of Protein Thermostability in Ionic Liquids. *Chem. Commun.,* 940–941.

Date submitted: 26th August 2005

Jeff D. Ballin, Ph.D.

Department of Biochemistry,
University of Maryland, Baltimore,
108 North Greene Street,
Baltimore, MD 21201, USA.
Tel: 410 706 7500 / 8903 Fax: 410 706 8297
jball003@umaryland.edu

Specialty Keywords: Nucleic acid interactions, RNA dynamics.

We study inter- and intramolecular interactions of nucleic acids and proteins. Current pursuits include the elucidation of mechanisms of mRNA turnover and the role of structural remodeling important for the association of protein factors involved in RNA decay regulation. To this end, we use time resolved fluorescence, single molecule FRET, anisotropy approaches.

-J.D. Ballin, I.A. Shkel, M.T. Record, Jr (2004). "Interactions of the KWK [6] Cationic Peptide with Short Nucleic Acid Oligomers: Demonstration of Large Coulombic End Effects on Binding at 0.1-0.2 M Salt," *Nucleic Acids Research,* 32, 3271-3281.

Date submitted: 6th April 2006

Aleksander Balter, Ph.D.

Institute of Physics,
N. Copernicus University,
Grudziadzka 5,
87-100 Torun, Poland.
Tel: +48 (56) 611 3216   Fax: +48 (56) 622 5397
balter@phys.uni.torun.pl

Specialty Keywords: Molecular biophysics, Photoluminescence, Sonoluminescence.

Dr. Balter's current research interests include: Surface enhanced fluorescence and Raman spectroscopy. Single bubble sonoluminescence.

-A. Kamińska, M. Kowalska and A. Balter (1999). A comparative study of the effect of exogenous and endogenous photostabilizers in the lens crystallin photodegradation, *J.Fluorescence* 9, 213-219.
-J. Szubiakowski, A. Balter, W. Nowak, K. Wisniewski and K. Aleksandrzak (1999) Substituent-sensitive anisotropic rotations of 9-acetoxy-10-phenylanthracenes. Fluorescence anisotropy decay and quantum-mechanical study, *Chem. Phys. Lett.* 313, 473-483.

Date submitted: 9th June 2005

Susan L. Bane, Ph.D.

Department of Chemistry,
State University of New York at Binghamton,
Binghamton, New York 13902,
USA.
Tel: 607 777 2927   Fax: 607 777 4478
sbane@binghamton.edu
chemistry.binghamton.edu/BANE/bane.html

Specialty Keywords: Microtubules, Ligand / Receptor interactions, New fluorescent probes.
AIM 2004 = 14.7

We are interested in determining the molecular mechanisms by which antimicrotubule drugs (such as paclitaxel (Taxol), colchicine, vinblastine, and combretastatin) interact with the protein tubulin and with microtubules. We use a variety of fluorescence spectroscopy techniques to elucidate these mechanisms. Design and synthesis of new fluorescent probes is also in progress.

-T. Ganesh, J. K. Schilling, R. K. Palakodety, R. Ravindra, N. Shanker, S. Bane, D. G. I. Kingston (2003). Synthesis and biological evaluation of fluorescently labeled epothilone analogs for tubulin binding studies *Tetrahedron* 59, 9979-9984.

Date submitted: 17th July 2005

Beniamino F. Barbieri, Ph.D.

ISS,
1602 Newton Drive,
Champaign, IL 61822-1061,
USA.
Tel: 217 359 8681   Fax: 217 359 7879
bb@iss.com
www.iss.com

Specialty Keywords: Fluorescence Instrumentation, Fluorescence Correlation Spectroscopy, Confocal Imaging.
As President of ISS, I am fostering our company's efforts and mission towards the development of innovative research-grade instrumentation, which will enable scientists to fully utilize the potentiality of fluorescence techniques in basic research. A parallel mission of our company is the development of novel medical instrumentation utilizing photonics tools. In our constant pursuit of innovations, ISS is wholly committed to offering quality and value added products and services that meet the present and future needs of our customers.

-Beniamino Barbieri, Ewald Terpetschnig and David M. Jameson; Frequency-Domain Fluorescence Spectroscopy Using 280-nm and 300-nm LEDs: Measurement of Proteins and Protein-Related Fluorophores; Analytical Biochemistry, in press.

Date submitted: Editor Retained.

Elisabeth Bardez, Ph.D.

Conservatoire National des Arts et Métiers,
292 rue Saint-Martin,
5141 Paris Cedex 03,
France.
Tel: +33 (0)1 40 27 2592   Fax: +33 (0)1 40 27 2362
bardez@cnam.fr

Specialty Keywords: Excited-state Proton Transfer, Fluorescent Sensors for Aluminum(III), Photoinduced Tautomerization.

Current interests: Photoinduced tautomerization in amphoterous bifunctional compounds (hydroxyquinolines, hydroxycoumarins). Photoinduced proton ejection from dihydroxynaphthalenes. Design of hexadentate fluorogenic ligands for aluminum determination includind bidentate sub-units as 8-hydroxyquinoline, chromotropic acid, etc.

-E. Bardez et al. (2001). From 8-hydroxy-5-sulfoquinoline to new related fluorogenic ligands for complexation of aluminium(III) and gallium(III). *New J. Chem.* 25, 1269 - 1280.

-E. Bardez (1999). Excited-state proton transfer in bifunctional compounds *Israel J. Chem.* 39, 319 - 332.

Date submitted: 18th July 2005

B. George Barisas, Ph.D.

Department of Chemistry,
Colorado State University,
Fort Collins, CO 80523,
USA.
Tel: 970 491 6641   Fax: 970 491 1801
barisas@lamar.colostate.edu

Specialty Keywords: Cell, Membrane, Dynamics, Diffusion, Lateral, Rotation, Phosphorescence, FRET, FRAP.

We examine the dynamics and distributions of cell surface molecules in relation to membrane signal transduction events in cells of the immune system and in gonadotropin-responsive cells. We measure lateral motions through photobleaching recovery and single-particle tracking, rotational motions through time-resolved phosphorescence anisotropy and fluorescence depletion anisotropy and spatial distributions through fluorescence resonant energy transfer and photoproximity labeling. We have developed new or improved implementations of each of the above techniques.

Date submitted: 26th September 2006

Mukulesh Baruah, Ph.D.

Chemistry Department, K.U.Leuven,
Celestijnenlaan 200F, Heverlee,
Leuven, 3001,
Belgium.
Tel: +32 (0)1 632 7398
Mukulesh.Baruah@chem.kuleuven.be &
mukulesh@yahoo.com
www.chem.kuleuven.be/research/mds/mds_en.html

Specialty Keywords: Synthetic Organic Chemistry, Fluorescent Probes, Enzymology.

Dr. Baruah's current research interests include: Synthesis of fluorescent ion probes. Synthesis of fluorogenic enzyme substrate and the studies of specific enzyme activities at single molecule level. Fluorescent labeling of proteins and oligosaccharides.

Date submitted: 8th July 2005

Wolfgang Becker, Ph.D.

Becker & Hickl GmbH.,
Nahmitzer Damm 30,
Berlin, 12277,
Germany.
Tel: 49 30 787 5632   Fax: 49 30 787 5734
becker@becker-hickl.de
www.becker-hickl.com

Specialty Keywords: TCSPC, FLIM, Time-resolved spectroscopy.

W. B. is a specialist of optical short-time measurement techniques. Since 1993 he is the head of Becker & Hickl GmbH in Berlin. His field of interest is development and application of Time-Correlated Single Photon Counting techniques. He likes cats, skiing and beach volleyball.

-W. Becker, H. Hickl, C. Zander, K.H. Drexhage, M. Sauer, S. Siebert, J. Wolfrum, Time-resolved detection and identification of single analyte molecules in microcapillaries by time-correlated single photon counting. Rev. Sci. Instrum. 70 (1999) 1835-1841.

-Wolfgang Becker, Axel Bergmann, Christoph Biskup, Thomas Zimmer, Nikolaj Klöcker, Klaus Benndorf, Multi-wavelength TCSPC lifetime imaging. Proc. SPIE 4620 (2002) 79-84.

Date submitted: 26th September 2006

Kevin D. Belfield, Ph.D.

Dept. of Chemistry & CREOL: College of Optics & Photonics,
University of Central Florida,
P.O. Box 162366, Orlando,
FL 32816-2366, USA.
Tel: 407 823 1028   Fax: 407 823 2252
kbelfiel@mail.ucf.edu
www.cas.ucf.edu/chemistry/faculty_belfield.php

Specialty Keywords: Multiphoton, Fluorescence Anisotropy, Two-Photon Fluorescent Probes.

The design and synthesis of well-defined organic materials with specific functionality and multiphoton-induced photochemistry and photophysics are of particular interest. With this understanding, applications that require 3D spatial control of photochemical reactions are being explored in optical data storage, bio-imaging, and photodynamic therapy.

-K. D. Belfield *et al.* (2006) Two-photon absorption of a supramolecular psuedoisocyanine J-aggregate assembly *Chem. Phys.* 320, 118-124.
-K. D. Belfield *et al.* (2005) Fluorene-based fluorescent probes with high two-photon action cross-sections for biological multiphoton imaging applications *J. Biomed. Opt. 10(5)*, 051402-1-051402-8.

Date submitted: 26th September 2006

Mário N. Berberan-Santos, Ph.D.

Centro de Química-Física Molecular,
Instituto Superior Técnico,
1049-001 Lisboa,
Portugal.
Tel: +351 21 841 9254   Fax: +351 21 846 4455
berberan@ist.utl.pt
dequim.ist.utl.pt/docentes/2219/english

Specialty Keywords: Photophysical Kinetics, Resonance Energy Transfer, Multichromophoric Systems.

Current interests: Photophysics of fullerenes and of multichromophoric systems. Radiative transport in atomic vapors and in scattering media.

-A.A. Fedorov, S.P. Barbosa, M.N. Berberan-Santos, Radiation propagation time broadening of the instrument response function in time-resolved fluorescence spectroscopy, *Chemical Physics Letters* 421 (2006) 157.

-M.N. Berberan-Santos, B. Valeur, Luminescence decays with underlying distributions: General properties and analysis with mathematical functions, *Journal of Luminescence*, in press

Date submitted: 8th July 2005

Axel Bergmann, Ph.D.

Becker & Hickl GmbH.,
Nahmitzer Damm 30,
Berlin, 12277,
Germany.
Tel: +49 30 787 5632   Fax: +49 30 787 5734
bergmann@becker-hickl.de
www.becker-hickl.com

Specialty Keywords: FLIM, TCSPC, Lifetime Analysis.

Axel Bergmann received his doctorate in physics from the Technical University of Berlin. He came to Becker & Hickl in 2000 and started to develop leading-edge hard and software products for photon counting instrumentation. As a scientific coworker within this company he is presently involved in the development of SPC-Image - an analysis tool which allows to create color-coded lifetime images from multidimensional TCSPC data. Research interests include the application of the FRET effect to biological systems by means of fluorescence lifetime imaging microscopy. During the last two years he had several publications about this topic in refereed scientific journals.

-R. Duncan, A. Bergmann, M.A. Cousin, D.K. Apps & M.J. Shipston, Multi-dimensional time-correlated single photon counting (TCSPC) fluorescence lifetime imaging microscopy (FLIM) to detect FRET in cells. Journal of Microscopy, Vol. 215, pp.1-12, 2003.

Date submitted: 23rd May 2006

Kankan Bhattacharyya, Ph.D.

Department of Physical Chemistry,
Indian Association for the Cultivation of Science,
Jadavpur, Kolkata 700 032,
India.
Tel: (91) 332 473 3542   Fax: (91) 332 473 2805
pckb@mahendra.iacs.res.in
www.iacs.res.in/pckb.html

Specialty Keywords: Ultra fast dynamics, Organized assembly.

Our major interest is to study dynamics in organized assemblies using femtosecond and picosecond time resolved fluorescence spectroscopy. Solvation dynamics, proton/electron transfer, isomerization and orientational dynamics are found to be dramatically retarded in protein, micelles & other organized assemblies. Most recently, we have showed that solvation dynamics of water at the active site of an enzyme and also in a partially folded protein display a component 100-1000 times slower than that in bulk water.

-K. Bhattacharyya, et al. (2006) *J. Phys. Chem. B* 110, 1056.
-K. Bhattacharyya et al. (2005) *Biochemistry* 44, 8940.

Date submitted: Editor Retained.

John J. Birmingham, Ph.D.

Unilever Research Port Sunlight,
Quarry Road East, Bebington,
Wirral, Merseyside, CH63 3JW,
United Kingdom.
Tel: +44 (0) 151 641 3351   Fax: +44 (0) 151 641 1841
John.Birmingham@unilever.com

Specialty Keywords: Photobleaching, Lifetime Imaging.
Research emphasis on development of fluorescence technologies to aid detection and imaging of industrially relevant ingredients deposited on both natural and man-made surfaces at low levels from consumer products. Key techniques include fluorescence photobleaching methods (time and frequency domains) and nanosecond timescale lifetime imaging, the latter implemented in the frequency domain for both widefield imaging and laser scanning geometries to suit a range of distance scales from microscopic to large macroscopic.
-J.J.Birmingham (1997) J.Fluorescence 7(1):45-54.
-J.J.Birmingham (1999) in A.Kotyk (ed) , Fluorescence Microscopy and Fluorescent Probes 3, Espero, Prague, pp.23-35.
-J.J.Birmingham (2002) in R.Kraayenhof (ed), Fluorescence Spectroscopy, Imaging and Probes, Springer, pp.297-316.

Date submitted: 8th July 2005

Giovanni Luca Biscotti, Ph.D.

Becker & Hickl GmbH.,
Nahmitzer Damm 30,
Berlin, 12277,
Germany.
Tel: +49 30 787 5632   Fax: +49 30 787 5734
biscotti@becker-hickl.de
www.becker-hickl.com

Specialty Keywords: TCSPC, FLIM, Time-resolved
Spectroscopy.

Scientific coworker for research and development of leading photon counting instrumentation at Becker & Hickl, Berlin. His university background provides a wide competence in Optics and Electronics environment. He supplies support on all elements involved in Fluorescence applications (from TCSPC basic principles to experiment devices).
-W. Becker, A. Bergmann, G. Biscotti, K. König, I. Riemann, L. Kelbauskas, C. Biskup
High-Speed FLIM Data Acquisiton by Time-Correlated Single Photon Counting. Proc SPIE 5232 (2004).
-W. Becker, A. Bergmann, G. Biscotti, A. Rück, Advanced time-correlated Single Photon Counting Technique. Proc. SPIE 5340 (2004).

Date submitted: 16th August 2006

Elaine Blackwood, Ph.D.

Edinburgh Instruments Ltd.,
2 Bain Square,
Livingston EH54 7DQ,
Scotland, UK.
Tel: +44 (0) 150 642 5300   Fax: +44 (0) 150 642 5320
elaine.blackwood@edinst.com
www.edinburghinstruments.com & www.edinst.com

Specialty Keywords: Fluorescence Spectrometers, Single
Photon Counting, Instrumentation.

Dr. Blackwood has been active in the field of fluorescence since 1989, Elaine joined Edinburgh Instruments in 1997, initially as an Applications Engineer for Analytical Division. In 1999 she progressed into a role as an installation engineer, and in the summer of 2000 moved into the sales team and has since worked on the global sales of the Analytical Division.

Date submitted: Editor Retained.   **Piotr Bojarski, Ph.D.**

Institute of Experimental Physics,
University of Gdańsk,
Wita Stwosza 57, Gdańsk,
80-952, Poland.
Tel: +48 58 552 9244
fizpb@univ.gda.pl

Specialty Keywords: Energy Transport, Aggregation, Monte - Carlo Simulation.

Areas of expertise: Multistep excitation energy transport and its trapping in disordered and ordered media, forward and reverse energy transfer, intermolecular aggregation, rotational depolarization of fluorescence, Kennard - Stepanov relation, excited state dipole moments, steady - state and time resolved fluorescence measurements, Monte - Carlo simulation.

-P. Bojarski, A. Kamińska , L. Kułak and M. Sadownik, Chem. Phys. Lett. (2003) 375, 547-552.

-P. Bojarski, L. Kułak, C. Bojarski and A. Kawski, J. Fluorescence (1995) 5 , 307 - 319.

Date submitted: 15th June 2005   **Guido Böse, Ph.D.**

Medical Physics and Biophysics,
& Center of Nanotechnology,
Gievenbecker Weg 11,
Münster, 48149, Germany.
Tel: +49 251 836 3825
gboese@uni-muenster.de

Specialty Keywords: Nanopores, 4Pi, FCS, Nuclear Pore.

The nanotechnological application of biological nanopores as molecular machines is a challenging goal of biomedical research. Using single molecule techniques as 4Pi-microscopy, Fluorescence Correlation Spectroscopy, single molecule tracking and electrical recording we characterize biological membrane pores with regard to biotechnological usage.

-R. Hemmler, G. Boese, R. Wagner und R. Peters. (2005). *Biophys. J.* 88:4000-4007.

-G. Böse, P. Schwille and T. Lamparter. (2004). *Biophysical J.* 87: 2013-2021.

Date submitted: 13th October 2006

Denis Boudreau, Ph.D.

Department of Chemistry &
Centre d'optique, photonique et laser (COPL),
Université Laval, Québec City,
PQ, Canada, G1K 7P4.
Tel: (418) 656 3287   Fax: (418) 656 7916
denis.boudreau@chm.ulaval.ca

Specialty Keywords: Novel Optical Biosensors
AIM 2006 = 16.8

Fluorescence in analytical and bioanalytical chemistry: development of new analytical techniques and biosensors based on fluorescence detection. Use of various laser-induced fluorescence techniques for the development of new analytical techniques; remote sensing of atmospheric pollutants using ultrafast laser-induced fractionation and excitation; characterization of DNA and protein optical biosensors based on FRET and superlighting.

-Doré K, Leclerc M, Boudreau D (2006) Investigation of a fluorescence signal amplification mechanism used for the direct molecular detection of nucleic acids. J Fluoresc 16:259-265.

Date submitted: 11th April 2006

Rebecca A. Bozym, (Ph.D. Student)

Department of Biochemistry and Molecular Biology,
University of Maryland School of Medicine,
108 N. Greene St, Baltimore,
MD 21201, USA.
Tel: 410 706 2588
Rbozy001@umaryland.edu

Specialty Keywords: Carbonic Anhydrase, Biosensor, Zinc.

My thesis focuses on the use of human carbonic anhydrase II as a signal transducer for detecting the level of free zinc in cells using a FRET-based ratiometric approach. Carbonic anhydrase is very selective for zinc over calcium and magnesium. Dapoxyl sulfonamide, an inhibitor of CA, binds only in the presence of zinc in the active site. We are continuing to monitor changes in intracellular zinc under apoptotic conditions.
-R.A. Bozym, R.B. Thompson, A.K. Stoddard, and C.A. Fierke (2006). Measuring Picomolar Intracellular Exchangeable Zinc in PC-12 cells using a Ratiometric Fluorescence Biosensor ACS Chem. Biol. 1(2), 103-111.
-R.B. Thompson, M.L. Cramer, R. Bozym, and C.A. Fierke (2002). Excitation Ratiometric fluorescence biosensor for zinc ion at picomolar levels J. Bio. Optics 7(4), 555-560.

Brand, L.
Bright, F. V.

Date submitted: Editor Retained.

Ludwig Brand, Ph.D.

Biology Department, Johns Hopkins University,
3400 North Charles Street,
Baltimore, MD 21218,
USA.
Tel: 410 515 7298   Fax: 410 516 7298
Ludwig.Brand@jhu.edu

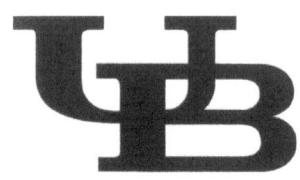

Specialty Keywords: Fluorescence, Proteins, Membranes.

The interest of our laboratory is to understand the static and dynamic structure of proteins, biological membranes, and nucleic acids. The work includes studies of the interactions between macromolecules and the relation between structure and function. A variety of excited-state processes such as proton transfer, energy transfer, exciplex and excimer formation and solvent relaxation are being investigated so that these processes can be better used to study biological macromolecules in vivo and in vitro.

-Toptygin, D. Savichenko, R.S., Meadow, N.D. and Brand, L.,"Homogeneous Spectrally and Time-Resolved Fluorescence Emission from Single-Tryptophan of IIA^{Glc} Protein.", Journal of Physical Chemistry B, 105, 2043-2055 (2001).

Date submitted: 24th June 2005

Frank V. Bright, Ph.D.

Department of Chemistry, University at Buffalo,
The State University of New York,
511 Natural Sciences Complex,
Buffalo, New York, 14260-3000, USA.
Tel: 716 645 6800 x 2162
chefvb@buffalo.edu
www.chem.buffalo.edu/bright.php

Specialty Keywords: Sensors, Materials, Benign Solvents.
AIM 2004 = 35.0

Research areas: [i] proteins in restricted environments; [ii] xerogel-based materials; [iii] microheterogeneous systems; [iv] environmentally benign solvents; [v] chemical analysis of things as they are; and [vi] laser-based instrumentation.

-R.M. Bukowski, R. Ciriminna, M. Pagliaro and F.V. Bright (2005). "High performance quenchometric oxygen sensors based on fluorinated xerogels doped with [Ru(dpp)$_3$]$^{2+}$," *Anal. Chem.* 77, 2670-72.

-G.A. Baker, S.N. Baker, S. Pandey and F.V. Bright (2005). "An analytical view of ionic liquids," *Analyst* 130, 800-8.

Date submitted: 9th August 2006   **Harry G. Brittain, Ph.D.**

Center for Pharmaceutical Physics,
10 Charles Road,
Milford, New Jersey 08848,
USA.
Tel: 908 996 3509   Fax: 908 996 3560
hbrittain@earthlink.net   hbrittain@centerpharmphysics.com

Specialty Keywords: Fluorescence of Organic Materials in the Solid State.

One area of interest concerns the effect of crystallographic polymorphism on solid-state fluorescence, and how spectroscopic results can be used to deduce information about the solid-state photophysics as affected by the variations in crystal structures.   In addition, the fluorescence of organic salts in the solid state is of interest as a means to understand the effects of protonation on the photophysics.   In those instances where the anion and cation are both separately fluorescent, the consequences of energy transfer are of great interest.

-H.G. Brittain, B.J. Elder, P.K. Isbester, and A.H. Salerno, "Solid-State Fluorescence Studies of Some Polymorphs of Diflunisal", *Pharm. Res.*, 22, 999-1006 (2005).
-H.G. Brittain, "Solid-State Fluorescence of the Trihydrate Phases of Ampicillin and Amoxicillin", *AAPS PharmSciTech*, 6(3), paper #55, 444-448 (2005).

Date submitted: 4th April 2006   **Rasmus Bro, Ph.D.**

Department of Food Science,
The Royal Veterinary and Agricultural University,
Rolighedsvej 30, Frederiksberg,
1958 Denmark.
Tel: 453 528 3296
rb@ kvl.dk
www.models.kvl.dk/users/rasmus

Specialty Keywords: PARAFAC, Multi-way, Chemometrics.

Works on making mathematical models for EEM's that enable resolving underlying fluorophores of complex EEM's of food, feed, blood, process-streams, environmental samples etc. The PARAFAC model is a wonderful tool originally developed in psychology, but perfectly suited for performing "mathematical chromatography": directly from EMM's of mixtures the pure component spectra and concentrations can be determined. On the above web, there are links to free programs, courses, literature etc.

Monograph: www.mli.kvl.dk/staff/foodtech/brothesis.pdf

Date submitted: Editor Retained.   **Jean-Claude Brochon, Ph.D.**

L.B.P.A., Ecole Normale Supérieure de Cachan, C.N.R.S.,
61, avenue du Président Wilson,
94235 Cachan Cedex,
France.
Tel: +33 (0) 1 47 40 2717   Fax: +33 (0) 1 47 40 2479
brochon@lbpa.ens-cachan.fr
www.lbpa.ens-cachan.fr/photobm/

Specialty Keywords: Proteins, Time-resolved Anisotropy, Data Analysis.

Structural dynamics and function of biological macromolecules from time-resolved fluorescence *in vitro*. Currently, protein dynamics, self-assembly of proteins, protein-nucleic acids and protein-protein interactions. A recent project in my laboratory is to extend these studies, *in vivo*, in using 2-photons confocal microscopy and FLIM techniques; application to retrovirus replication. High hydrostatic pressure for study of protein plasticity. Application of the Maximum Entropy Method of data analysis in time-resolved spectroscopies.

-Deprez, E., Tauc, P., Leh, H., Mouscadet, J-F., Auclair, C. Hawkins, M. E., Brochon, J-C., DNA binding induces dissociation of the multimeric form of HIV-1 integrase : A time-resolved fluorescence anisotropy study, Proc. Nat. Acad. Sci. USA, (2001) 98, 10090- 10095.

Date submitted: 14th August 2005   **Fred A. M. Brouwer, Ph.D.**

Van't Hoff Institute for Molecular Sciences,
University of Amsterdam,
Nieuwe Achtergracht 129, 1018 WS Amsterdam,
The Netherlands.
Tel: +31 20 525 5491   Fax: +31 20 525 5670 / +31 84 871 0814
A.M.Brouwer@uva.nl
www.science.uva.nl/~fred

Specialty Keywords: Motor Molecules, (Micro) Photochemistry, Computational Chemistry, Fluorescent Probes.

Our main research theme is "motor molecules", synthetic analogs of motor proteins, that is: molecules that can be made to undergo large-amplitude motion. We mainly use photoinduced electron transfer and E-Z isomerization as stimuli.
Among other subjects of study are fluorescent and highly solvatochromic electron-donor acceptor molecules, which we apply as probes of the dynamics of solutions and polymer media.

-A. M. Brouwer, C. Frochot, F. Gatti, D. A. Leigh, L. Mottier, F. Paolucci, S. Roffia and G. W. H. Wurpel, (2001), *Science*, 291, 2124-2128.
-P.D. Zoon, A.M. Brouwer, *ChemPhysChem*, in press August 2005.

Date submitted: 30th October 2006 **Martha P. Brown, Ph.D.**

Biologics and Biomolecular Sciences,
Boehringer Ingelheim Pharmaceuticals, Inc.,
900 Ridgebury Road, Ridgefield,
CT 06877, USA.
Tel: 203 798 5756
mbrown2@rdg.boehringer-ingelheim.com

Specialty Keywords: uHTS, Anisotropy, Energy transfer.

My research applies fluorescence spectroscopy toward defining and characterizing the mechanism of action of drug candidates at the molecular level as well as to the development of ultra high throughput screening assays. The equilibrium binding properties of drug candidates with their protein targets are studied to understand target-compound interactions.

-M. August, L. Patnaude, J. Hopkins, J. Studts, E. Gautschi, A. Shrutkowski, A. Kronkaitis, M.P. Brown, A. Kabcenell, and D. Rajotte (2006), Development of a high-throughput assay to measure histidine decarboxylase activity, Jn. Biomol. Screen. 11(7), 816-821.

-A. Prokopowicz, M.P. Brown, J. Wildeson, S. Jakes and M. Labadia (2005) Fluorescent Probes for use in Protein Kinase Inhibitor Binding Assay, United States Patent Application Publication US2005/0100978 A1.

Date submitted: 22nd September 2006 **Charles K. Brush, Ph.D.**

Pierce Milwaukee, LLC.,
2202 North Bartlett Ave,
Milwaukee, WI 53202,
United States.
Tel: 001 414 227 3738 Fax: 001·414 227 3774
Chuck.brush@perbio.com
www.piercebc.com/

Specialty Keywords: Carbocyanine, Nucleic Acids,
Monomethine.

Dr. Brush is interested in developing new fluorescent dyes and methods for labeling and detection of nucleic acids, either by conjugation as phosphoramidites or reactive esters, by incorporation of labeled NTPs, or by non-covalent association. Of particular current interest are carbocyanine, styryl, and monomethine dyes for selective labeling in molecular biology assays.

-Brush CK and Anderson ED, Indocarbocyanine-linked Phosphoramidites, US Patent 5,556,959, Sept. 17, 1996.

-Brush CK Fluorescein Labeled Phosphoramidites, US Patent 5,583,236, Dec. 10, 1996.

Date submitted: 16th August 2005

Srinivasa Buddhdu, Ph.D.

Department of Physics,
Sri Venkateswara University,
Tirupati 517 502,
India.
Tel: 91 877 226 1611   Fax: 91 877 226 1611
sbuddhudu@hotmail.com
drsb99@hotmail.com

Specialty Keywords: Luminescence, Glasses, Polymers.

Main research interests are focused on a variety of luminescent and opto-elelctronic materials (glasses, phosphors, crystals, ceramics, complexes, compounds, polymers, powder samples, thin-films and other related inorganic/organic compounds) containing rare-earth (4f) ions and also transition metal ions (3d) in evaluating their optical performance from absorption & luminescence, energy-transfer fluorescence and up-conversion emission processes and mechanisms involved.

-M. Morita, S. Buddhudu, D. Rau and S. Murakami (2004). Photoluminescence and Excitation Energy Transfer of Rare Earth Ions in Nanoporous Xerogel and Sol-Gel SiO_2 Glasses Structure & Bonding 107, 115-143

Date submitted: 22nd September 2006

Peter J. Butler, Ph.D.

Bioengineering,
The Pennsylvania State University,
228 Hallowell Building, University Park,
Centre County, 16802, USA.
Tel: 814 865 8086   Fax: 814 863 0490
pjbbio@engr.psu.edu
www.bioe.psu.edu

Specialty Keywords: Vascular Biology, Endothelial Cells,
Mechanotransduction, Spectroscopy, Microscopy.

Our laboratory is interested in applying multimodal imaging techniques to study the effects of mechanical forces (e.g. fluid shear stress) on the dynamics of molecules in living cells and tissues involved in mechanotransduction. We employ confocal microscopy, time correlated, single photon counting spectroscopy, laser trapping, TIR and other modalities simultaneously on the same microscope to study the spatial and temporal aspects of mechano-activation of cells. We wish to use these techniques to understand the molecular bases of mechanically-induced changes in vascular biology.

Date submitted: 15th August 2005

Patrik R. Callis, Ph.D.

Department of Chemistry and Biochemistry,
Montana State University,
Bozeman, MT 59717,
USA.
Tel: 406 994 5414   Fax: 406 994 5407
pcallis@montana.edu
chemistry.montana.edu/callis.html/

Predicted Tryptophan Fluorescence Quantum Yields

Specialty Keywords: Proteins, Quenching, Electron Transfer.

Building on years of experimental and theoretical work on the excited states of tryptophan, we now are focused on hybrid quantum mechanical/molecular dynamics computations of photoinduced electron transfer rates in proteins. The current methods are useful for predicting tryptophan fluorescence quantum yields, and progress is being made towards the ability to predict quenching of flavin and dye fluorescence *by* tryptophan and tyrosine.

-P. R. Callis and T. Liu, (2004). Quantitative predictions of fluorescence quantum yields for tryptophan in proteins. *J. Phys. Chem. B.* 104(14), 4248-4259.
-T. Liu, P. R. Callis, B. H. Hesp, M. de Groot, W. J. Buma, and J. Broos (2005). Ionization potentials of fluoroindoles and the origin of non-exponential tryptophan fluorescence decay in proteins. *J. Am. Chem. Soc* 127(11), 4104-4113.

Date submitted: 25th September 2006

Haishi Cao, Ph.D.

Pacific Northwest National Laboratory,
Biological Science Division,
PO Box 999, MS P7-56, Richland,
WA 99354, USA.
Tel: 509 376 3180   Fax: 509 376 6767
haishi.cao@pnl.gov

Specialty Keywords: Fluorescent Probe, Carbohydrate, Protein.
AIM 2006 = 4.4

My research interests are being investigated in two main areas. Fluorescent chemosensors for carbohydrate recognition and development of cell permeable multiuse affinity probes for protein. The underlying theme is organic synthesis. In most of case probes will be utilized for real-time imaging in living cell.

-H. Cao, B. Chen, T. C. Squier, M. U. Mayer, (2006) " CrAsH: A biarsenical multi-use affinity probe with low non-specific fluorescence" *Chem. Commun.* 2601 – 2603.
-H. Cao, V. Chang, R. Hernandez, M. Heagy (2005). Matrix library screening of substituted *N*-Aryl-1,8-naphthalimides reveals new dual fluorescent dyes and unusually bright pyridine derivatives. *J. Org. Chem.* (70), 4929-4934.

Date submitted: 22nd September 2006   **Elisabete M. Castanheira, Ph.D.**

Departamento de Física,
Universidade do Minho,
Campus de Gualtar, 4710-057 Braga,
Portugal.
Tel: + 351 25 360 4321   Fax: +351 25 360 4061
ecoutinho@fisica.uminho.pt
www.fisica.uminho.pt

Specialty Keywords: Molecular Spectroscopy, Biophysics, Microheterogeneous Systems.

Current interests: Fluorescent probes; biological membranes; self-assembly molecules; new fluorescent drugs; Langmuir-Blodgett films; molecular spectroscopy; kinetics.

-M.-J.R.P. Queiroz, E.M.S. Castanheira, A.M.R. Pinto, I.C.F.R. Ferreira, A. Begouin, G. Kirsch (2006), Synthesis of the first thieno-δ-carboline. Fluorescence studies in solution and in lipid vesicles, *J. Photochem. Photobiol. A: Chem.* 181 (2006) 290-296.

-G. Hungerford, A.L.F. Baptista, P.J.G. Coutinho, E.M.S. Castanheira, M.E.C.D. Real Oliveira (2006), Interaction of DODAB with neutral phospholipids and cholesterol studied using fluorescence anisotropy, *J. Photochem. Photobiol. A: Chem.* 181 (2006) 99-105.

Date submitted: 1st April 2005   **Zoran G. Cerovic, D.Sc.**

Biospectroscopy – ESE-CNRS,
Bât 362, Orsay,
F-91405,
France.
Tel: +33 (0)16 446 8209   Fax: +33 (0)16 915 7238
zoran.cerovic@ese.u-psud.fr
www.ese.u-psud.fr

Specialty Keywords: Chlorophyll, Polyphenols, Optical Sensors.

AIM 2004 = 4.2

Investigations on the origin of variable chlorophyll fluorescence *in vivo*. Time–resolved measurements of fluorescence. Investigation on the origin of blue-green fluorescence of plants, and on the UV-excited fluorescence of leaves in general. Design of fluorescence signatures for remote sensing of vegetation. Design of optical sensors for sustainable agriculture based on plant autofluorescence.

-Cerovic, Z. G. et al. (2002), The use of chlorophyll fluorescence excitation spectra for the nondestructive in situ assessment of UV–absorbing compounds in leaves, *Plant Cell Environ.*, 25:1663-1676.

-Goulas, Y. et al. (2004), Dualex: A new instrument for field measurements of epidermal UV-absorbance by chlorophyll fluorescence, *Appl. Opt.*, 43:4488-4496.

Date submitted: Editor Retained.

Olga Nikolaevna Chaikovskaya, Ph.D.

Siberian Physical Technical Institute,
Novo-Sobornaya, Tomsk,
634050,
Russia.
Tel: +7 (3822) 53 3426   Fax: +7 (3822) 53 3034
tchon@phys.tsu.ru

Specialty Keywords. Photochemistry, Fluorescent
Spectroscopy, Photolysis.

The method of fluorescent spectroscopy is used to investigate the influence of the pH of the medium and of the exciting radiation wavelength on phototransformations of *o*- and *p*-cresol in water under UV irradiation. It is demonstrated that the efficiency of cresol photodecomposition decreases with the increasing pH of the medium. The efficiency of cresol phototransformations in an alkaline medium is higher under irradiation at 283 nm, whereas in a neutral medium, it is higher under irradiation at 222 nm.

-A. Svetlichnyi, O. N. Chaikovskaya, O. K. Bazyl', *et al.* (2001). *High-Energy Chemistry*, 35 258 (Translated from Khimiya Vysokikh Energii, Russia).

Date submitted: 31st July 2006

Abhijit Chakrabarti, Ph.D.

Biophysics Division,
Saha Institute of Nuclear Physics,
1/AF Salt lake, Kolkata,
700064, India.
Fax: 91 33 233 7463
abhijit.chakrabarti@saha.ac.in

Specialty Keywords: Spectrin, Erythrocyte Membrane.

Present research of my group interest is on protein-protein cross talk and the role of the cell surface membrane lipids in red cell disorders e.g. hemoglobinopathy. We have been working on various aspects of spectrin-based network and hemoglobin variants and their effects on oxidative cross-linking, membrane asymmetry, skeletal architecture and protein profiles of the red cells implicated in thalassemia and leukemia.

-Conformational study of spectrin in presence of submolar concentrations of denaturants. (2005). J Fluorescence 15, 61-70.

-Devaki A Kelkar, Amitabha Chattopadhyay, *Abhijit Chakrabarti* and Malyasri Bhattacharyya. Effect of ionic strength on the organization and dynamics of tryptophan residues in erythroid spectrin : A fluorescence approach. (2005). Biopolymers, 77, 325-334.

Date submitted: 5th April 2006

Philip J. Chan, Ph.D.

Departments of Gyn / Ob – IVF, Physiology & Pharmacology,
Loma Linda University School of Medicine,
11370 Anderson Street, Suite 3950,
Loma Linda, California 92354, USA.
Fax: 909 558 2450
pchann@yahoo.com
www.llu.edu/lluhc/fertility

Specialty Keywords: Infertility, Andrology, Embryology.

The research interests center on the role of papillomavirus in gametes, embryos, the characteristics of transgenic sperm, and mutations in proto-oncogenes and BRCA1. In addition, the research extends to studying fluorescent assays based on dyes and nanoparticles / q-dots for cell apoptosis and preimplantation analyses.

-Bosman SJ, Nieto S, Patton WC, Jacobson JD, Corselli J, Chan PJ. Development of mammalian embryos exposed to mixed-size nanoparticles. Clin. Exp. Obstet Gynecol 32:222-224 (2005).

-Chan, P.J., Jacobson, J.D., Corselli, J., Patton, W.C. A simple zeta method for sperm selection based on membrane charge. Fertil Steril 85:481-486 (2006).

Date submitted: 18th October 2006

Lin L. Chandler, Ph.D.

Fluorescence Division,
HORIBA Jobin Yvon,
3880 Park Avenue,
Edison, NJ 08820-3012, USA.
Tel: 732 494 8660   Ext: 236   Fax: 732 549 5157
Lin_Chandler@jyhoriba.com
www.jobinyvon.com

Specialty Keywords: UV Microscopy, Anisotropy, Frequency-domain.

Dr. Chandler is a member of a team of scientists providing fluorescence applications support, training and new methods development for users of HORIBA Jobin Yvon's spectrofluorometers. Support is provided for all users interested in applying high sensitivity photon-counting, steady-state fluorescence spectroscopy, fluorescence microscopy and picosecond time-resolved, frequency-domain methods to their own research projects.

Date submitted: 25th April 2006

Amitabha Chattopadhyay, Ph.D.

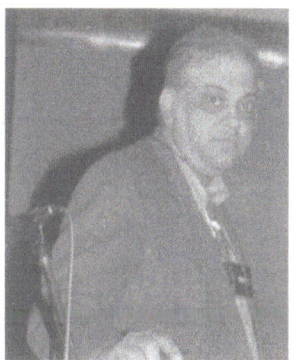

Centre for Cellular and Molecular Biology,
Uppal Road,
Hyderabad 500 007,
India.
Tel: +91 402 719 2578   Fax: +91 402 716 0311
amit@ccmb.res.in
www.ccmb.res.in

Specialty Keywords: Biomembranes and Other Organized Assemblies, Solvent Relaxation, FRAP.
AIM 2004 = 61.1

My major research interest is the application of fluorescence spectroscopic approaches to problems in membrane and receptor biology. We have successfully utilized approaches based on slow solvent relaxation rates in organized molecular assemblies such as membranes, micelles, and reverse micelles to address key questions related to their organization and dynamics including issues in membrane domains.   Another area of interest is the application of fluorescence techniques to explore organization and dynamics of the serotonin$_{1A}$ (5-HT$_{1A}$) receptor in order to understand its function and organization in the membrane.

-T.J. Pucadyil, S. Kalipatnapu, K.G. Harikumar, N. Rangaraj, S.S. Karnik, and A. Chattopadhyay (2004) *Biochemistry* 43: 15852-15862.
-H. Raghuraman and A. Chattopadhyay (2004) *Biophys. J.* 87: 2419-2432.

Date submitted: 8th September 2006

Alex F. Chen, Ph.D.

Departments of Pharmacology and Neurology,
Cell and Molecular Biology Prog, & Neuroscience Program,
Michigan State University, East Lansing,
MI 48824-1317,USA.
Tel: 517 432 2730   Fax: 517 353 8915
chenal@msu.edu

Specialty Keywords: Vascular biology and disease, Gene- and cell-based therapies.

The research interest of my laboratory is vascular biology and vascular disease. Specifically, we study vascular dysfunction and complications in hypertension, diabetes and stroke in rodent models. Our experimental approaches consist of *in vitro* biochemical, molecular biology and pharmacological techniques and *in vivo* pharmacological methods in animal models, including gene- and EPC cell-based therapies and fluorescent confocal microscopy techniques.

-Li LX et al. *Circulation* 107, 1053-1058, 2003.
-Luo JD et al. *Circulation*, 110, 1238-1245, 2004.

Date submitted: 29th July 2005   **Herbert C. Cheung, Ph.D.**

Department of Biochemistry,
University of Alabama at Birmingham,
1918 University Boulevard, Birmingham,
Alabama 35294-0005, U.S.A.
Fax: 205 934 2485
hccheung@uab.edu

Specialty Keywords: Cardiac Contractility, FRET.

My major interests are in mechanistic studies of activation of cardiac myofilaments by calcium and its regulation by myosin motor, phosphorylation of troponin subunits, and disease-related mutations in troponin. Current work is focused in distance mapping of the regulatory proteins by FRET. FRET-based stopped-flow is used to investigate the kinetics of conformational switching in the filament to understand the role of structural dynamics in contractility. We use experimental FRET distances, simulated annealing and molecular dynamics to construct "low resolution" molecular models of troponin and its complex for additional insights on the role of structure in cardiac muscle. This approach is not limited by the size of the proteins and does not require protein crystallization. We also use FRET to study conformational switching in kinesins and other proteins.

-W.-J. Dong, J.M. Robinson, J.Xing, and H.C. Cheung (2003). Kinetics of conformational switching in cardiac troponin determined by FRET *J. Biol. Chem.* 278, 42394-42402.

Date submitted: 25th August 2005   **Hung-Yoon Choi, Ph.D.**

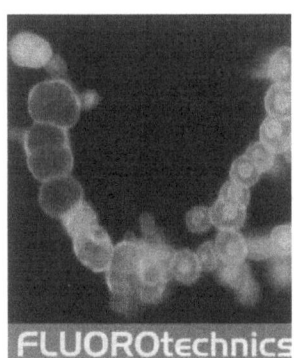

Fluorotechnics Pty Limited,
Dept. of Biological Sciences,
Macquarie University,
Sydney, Australia, 2109.
Tel: 61 (02)9 850 6267
ychoi@rna.bio.mq.edu.au
www.fluorotechnics.com

Specialty Keywords: Microbial Physiology and Genetics, Proteomics.

Hung-Yoon Choi's Research areas include: Microbial responses to environmental stresses, Induction of fungal secondary metabolites and Application of natural fluorophores to Genomics, Proteomics and Cellulomics.

-J.A. Mackintosh, H.-Y. Choi, S.-H. Bae, D. D. Van Dyk, B. C. Ferrari, N. M. Verrills, P. J. Bell, Y.-K. Paik, D. A. Veal and P. Karuso, (2003), A fluorescent natural product for ultrasensitive detection of proteins in 1-D and 2-D gel electrophoresis, *Proteomics* 3, 2273-88.
-H.-Y. Choi, M. Ryder, M. Gillings, H. Stokes, K. Ophel-Keller, and D. A. Veal, (2003), Survival of a *lacZY* marked strain of *Pseudomonas corrugate* following field release. *FEMS Microbiology Ecology*, 43, 367-374.

Date submitted: 25th September 2006

Mustafa H. Chowdhury, Ph.D.

Center for Fluorescence Spectroscopy,
Dept. of Biochemistry and Molecular Biology,
University of Maryland School of Medicine,
725 West Lombard St, Baltimore, Maryland, 21201, USA.
Tel: 410 706 7500
mustafa@cfs.umbi.umd.edu
cfs.umbi.umd.edu/cfs/CFShome.html
Specialty Keywords: MEF, SPCE, FDTD, Nanotechnology.
AIM 2006 = 13.1

Mustafa Chowdhury began his postdoctoral position at the Center for Fluorescence Spectroscopy on Sept. 2005. His research interest includes development of plasmonic nanostructures for modifying the photophysical properties of fluorophores using metal nanoparticles. He has experience in the fabrication and imaging of metallic nanstructures structures for plasmonic applications. He has experience in using computer algorithms to simulate the near-field and far-field optical properties of metallic nanostructures.

-M.H. Chowdhury, K. Aslan, S.N. Malyn, J.R. Lakowicz, C.D. Geddes, Metal-enhanced chemiluminescence: Radiating plasmons generated from chemically induced electronic excited states, *Applied Physics Letters* (2006) 88, 173104.

Date submitted: 25th May 2006

Rivka Cohen-Luria, Ph.D.

The Department of Chemistry,
Ben-Gurion University,
P.O. Box 653, Beer Sheva,
Israel, 84105.
Tel: 972 8 646 1191   Fax: 972 8 647 2943
riky@bgu.ac.il

Specialty Keywords: Protein-Protein, Protein-Lipid & Protein-Ligand / Drug Interactions.

Research topics: The role of hydrophobic interactions in membrane and non-membrane protein function and regulation, signal transduction, cell cycle and proliferation and intercellular interactions, angiogenesis, apoptosis, magnetic field effects on biological systems.
Crystallization and preliminary X-ray analysis of the apo form of *Escherichia coli* tryptophanase.

-Kogan A. , G.Y. Gdalevsky, R. Cohen-Luria, A.H. Parola and Y. Goldgur. *Acta Crystallogr. D Biol. Crystallogr.*, 60, 2073-5, 2004.
-Radical scavengers suppress low frequency EMF enhanced proliferation in cultured cells and stress effects in higher plants. A.H. Parola, D. Kost, G. Katsir, E. Ben-Izhak Monselise and R. Cohen-Luria. *The Environmentalist* 25, 103-111 (2005).

Comerford, J. J.
Coutinho, P. J. G.

Date submitted: 18 October 2006   **Jeffrey J. Comerford, Ph.D.**

FT-IR / Raman Product Marketing Manager,
Varian Australia Pty. Ltd.,
679 Springvale Road,
Mulgrave, 3170, Australia.
Tel: +61 39 566 1483   Fax: +61 39 566 1196
jeff.comerford@varianinc.com
www.varianinc.com

Specialty Keywords: Molecular Spectroscopy, Imaging,
Analytical Instrumentation, FT-IR.

My background is in molecular spectroscopy, in particular, the solution and photochemical behavior of square planar platinum (II) anti-cancer drugs. Experienced in the use of fluorescence, UV-Vis absorption and high pressure spectroscopy techniques, I previously was responsible for Varian's global fluorescence business, which included the Varian Cary Eclipse spectrophotometer. Following this, I spent a couple of years in the US in sales/sales support, my territory being the Western US region. In September 2004, Varian acquired a new product line in FT-IR spectrometers, of which I now have global responsibility in managing our infrared business.

Date submitted: 22nd Sept. 2006   **Paulo J. G. Coutinho, Ph.D.**

Departamento de Física,
Universidade do Minho,
Campus de Gualtar, 4710-057 Braga,
Portugal.
Tel: + 351 253 60 4321   Fax: +351 253 60 4061
pcoutinho@fisica.uminho.pt
www.fisica.uminho.pt

Specialty Keywords: Kinetics in confined media, Biophysics,
nanoparticles production by surfactant templating.

Current interests: Biophysics, kinetics in confined media, self-assembly molecules, computer simulations, solar energy conversion, dynamics in biological membranes, photodegradation of pollutants, semiconductor nanoparticles, Langmuir-Blodgett films, surfactant templating.

-P.J.G. Coutinho, M. Teresa C.M. Barbosa (2006), Characterization of TiO_2 Nanoparticles in Langmuir-Blodgett Films, *J. Fluorescence* 16 (2006) 387-392.

-G. Hungerford, A.L.F. Baptista, P.J.G. Coutinho, E.M.S. Castanheira, M.E.C.D. Real Oliveira (2006), Interaction of DODAB with Neutral Phospholipids and Cholesterol studied using Fluorescence Anisotropy, *J. Photochem. Photobiol. A: Chem.* 181 (2006) 99-105.

Date submitted: 26th September 2006

Peter Czerney, Ph.D.

Dyomics GmbH.,
Winzerlaer Str.2,
07745 Jena,
Germany.
Tel: +49 364 150 8200   Fax: +49 364 150 8201
p.czerney@dyomics.com
www.dyomics.com

Specialty Keywords: Customised Dyes.

Dr. Czerney is the Founder and Managing Director of Dyomics (Dyes and Genomics / Proteomics). Peter, earned his Ph.D. in Dye Chemistry from TU Dresden / Germany (1983).

The research interests include dye chemistry, new high quality "tailor made" dyes for bioanalysis and related fields of high technology, markers for biomolecular research and diode laser-compatible fluorescent labels.

Date submitted: 26th September 2006

Sabato (Tino) D'Auria, Ph.D.

Institute of Protein Biochemistry,
Italian National Research Council,
Via Pietro Castellino, 111,
80131 Napoli, Italy.
Tel: +39 081 613 2250   Fax: +39 081 613 2277
s.dauria@ibp.cnr.it
www.ibp.cnr.it

Specialty Keywords: Protein Fluorescence, Enzyme Fluorescence, Biosensors.

The research activity of Dr. D'Auria's lab is focused on the development of advanced fluorescence protein-based biosensors for analytes of high social interests. Dr. D'Auria's lab is well equipped for the study of protein structure and the development of protein sensors. In fact, the dr. D'Auria's lab has modern steady-state and time-resolved fluorometers, circular dichroism spectropolarimeter, Biacore and FCS instrumentation.

-Writing 3D protein nanopatterns onto a silicon nanosponge; S. Borini, Sabato D'Auria, M. Rossi, A. M. Rossi *Lab-on-a-Chip* (2005) Oct;5(10):1048-52
-Advanced Protein-Based Biosensors: Glucose Biosensors as a Model for Analyses of High Social Interest M. Staiano, P. Bazzicalupo, M. Rossi, and Sabato D'Auria; *Molecular BioSystems* (2005)Dec;1(5-6):354-62.

Date submitted: 05th April 2006

Robert E. Dale, Ph.D.

University of London

King's College London,
School of Biomedical & Health Sciences,
Randall Division of Cell & Molecular Biophysics,
3rd Floor, New Hunt's House, Guy's Hospital Campus,
London SE1 1UL, United Kingdom.
Tel: 44 (0) 207 848 6471   Fax: 44 (0) 207 848 6435
bob.dale@kcl.ac.uk

Specialty Keywords: Orientation, Depolarization, FRET.

Research interests include: Theory and practice of steady-state and time-resolved fluorescence and fluorescence polarization spectroscopy and Förster long-range resonance excitation energy transfer (FRET) as probes of molecular, macro-molecular and supra-molecular structure and dynamics in their relation to biochemical and biological function and mechanism. Recent efforts centre on muscle cross-bridge (lever arm) orientation and reorientational dynamics by fluorescence depolarization, and the identification of putative membrane rafts by FRET.

-M.A.Acasandrei, R.E.Dale, M.vandeVen & M.Ameloot, *Chem.Phys.Lett.* 419 (2006) 469-473.

Date submitted: 11th August 2005

Suresh Das, Ph.D.

Photosciences and Photonics,
Chemical Sciences Division,
Regional Research, Laboratory (CSIR),
Trivandrum-695 019, India.
Tel: 91 471 251 5318   Fax: 91 471 249 0186
sdaas@rediffmail.com

Specialty Keywords: Liquid Crystals, Luminescent Materials, Near Infrared Dyes.
AIM 2004 = 18.0

Research interests include a) design of novel photoresponsive liquid crystalline materials for imaging applications, b) understanding structure property relationships of solid state luminescent materials and c) synthesis of novel squaraine-based near infra red dyes with potential for applications as sensitizers for large band-gap semiconductors and photodynamic therapy, and as fluorescent probes.

-R. Davis, N. P. Rath, S. Das, (2004), *Chem. Commun.*, (1), 74-75.

-R. Davis, V. A. Mallia, S. Das and N. Tamaoki, (2004) *Adv. Funct. Mater.* 14 (8), 743-748.

Date submitted: Editor Retained.

Lesley Davenport, Ph.D.

Department of Chemistry, Brooklyn College of CUNY,
2900 Bedford Avenue, Brooklyn,
New York 11210,
USA.
Tel: 718 951 5750   Fax: 718 951 4827
LDvnport@brooklyn.cuny.edu
academic.brooklyn.cuny.edu/chem/davenport/

Specialty Keywords: Time-resolved Fluorescence, Lipid Packing and Dynamics, Fluorescent Probes.

Research in our laboratory is currently focused on employing fluorescence methods for studying molecular interactions. We are particularly interested in employing long-lived fluorescence probes for investigating submicrosecond dynamics.

-L. Davenport, B. Shen, T.W. Joseph and M.P. Straher (2001) A Novel Fluorescent Coronenyl-Phospholipid Analogue for Investigations of Submicrosecond Lipid Fluctuations. *Chem. Phys. Lipids*. 109, 145-156.
-P. Targowski and L. Davenport (1998) Pressure Effects of Submicrosecond Phospholipid Dynamics Using a Long-Lived Fluorescence Probe, *J. Fluorescence*, 8, 121-128.

Date submitted: 27th September 2006

Sander J. G. de Jong, M.Sc.

Lambert Instruments,
Turfweg 4,
9313 TH Leutingewolde,
The Netherlands.
Tel: +31 50 501 8461   Fax: +31 50 501 0034
jgsdejong@lambert-instruments.com
www.lambert-instruments.com

Specialty Keywords: Fluorescence, Frequency Domain FLIM, FRET, Acquisition / Analysis Software.

Currently, I am working on an improved version of our software package LI-FLIM that is used with the Lambert Instruments Fluorescence lifetime imaging Attachment (LIFA) to record and analyze fluorescence lifetime images.

One of our goals is to be able to do video rate lifetime imaging. This will enable you to see, in real-time, changes in the (spatial distribution of) fluorescence lifetimes, for example caused by FRET.

-K.W.J. Stoop, K. Jalink, S.J.G. de Jong, L.K. van Geest (2004) Measuring FRET in living cells with FLIM *Proceedings of 8th Chinese International Peptide Symposium in press.*
-K.W.J. Stoop, K. Jalink, S.J.G. de Jong, L.K. van Geest (2004) Measuring FRET in living cells with FLIM *Proceedings of 8th Chinese International Peptide Symposium.*

Date submitted: Editor Retained.

Frans C. De Schryver, Ph.D.

Departmentof Chemistry,
KU Leuven,
Celestijnenelaan 200F Heverlee,
B-3001, Belgium.
Tel: +321 632 7405   Fax: +321 632 7989
Frans.deschryver@chem.kuleuven.ac.be
www.chem.kuleuven.ac.be/research/mds/members.htm

Specialty Keywords: Time Resolved Fluorescence, Confocal Microscopy, Single Molecule Spectroscopy.

The research group has over the years established an ensemble of techniques with special emphasis on pico second fluorescence decay ac-quisition and analysis by self developed algorithms (global and compartmental analysis), up-conversion and single molecule spectroscopy. The group has set up tools to down scale in size and in time the object of the photochemical and photophysical study.

-K. Velonia, O. Flomen-bom, D. Loos, S. Masuo, M. Cotlet, Y. Engelborghs, J. Hofkens, A.E. Rowan, J. Klafter, R.J.M. Nolte, F.C. De Schryver Single enzyme kinetics of CALB catalyzed hydrolysis Angewandte Chemie, 43, 2-6 (2004).

Date submitted: 5th May 2006

Amilra P. de Silva, Ph.D.

School of Chemistry,
Queen's University,
Stranmillis Road, Belfast,
BT9 5AG, Northern Ireland.
Tel: + 44 289 097 4422   Fax: + 44 289 097 4890
a.desilva@qub.ac.uk
www.ch.qub.ac.uk/staff/desilva/apds.html

Specialty Keywords: Sensors, Molecular logic, Switches.
AIM 2004 = 23.5

We established the generality of the luminescent PET (photoinduced electron transfer) sensor/switch principle - one of the most popular sensor/switch designs and now used by many laboratories around the world. The first example of intrinsically molecular logic in the primary research literature came from our laboratories. We continue to develop these two lines.

S. Uchiyama, A.P. de Silva and K. Iwai (2006) Luminescent Molecular Thermometers *J. Chem. Educ*. 83, 720-727.

-D.C. Magri, G.J. Brown, G.D. McClean and A.P. de Silva (2006) Communicating Chemical Congregation: A Molecular AND Logic Gate with Three Chemical Inputs as a 'Lab-on-a-Molecule' Prototype *J. Am. Chem. Soc.* 128, 4950-4951.

Date submitted: 15th June 2004

Soma De, Ph.D.

Center for Neuropharmacology and Neuroscience,
Albany Medical Center,
47 New Scotland Avenue, Albany,
12208, NY, USA.
Tel: 518 262 5416   Fax: 518 262 4348
des@mail.amc.edu & des@mail.rockefeller.edu

Specialty Keywords: Age-related Macular Degeneration (AMD), Drusen, Membrane Mimics.

The formation of drusen and other deposits below the retinal pigment epithelial (RPE) cells on Bruch's membrane and accumulation of lipofuscin in RPE cells are initial steps in the pathogenesis of AMD. My research focuses on to understand the origin of drusen and how A2E, a component of lipofuscin, induces apoptosis of RPE cells. I have also applied fluorescence spectroscopy extensively to study the membrane properties of synthetic dimeric lipid vesicles.

-S. De and T. P. Sakmar, (2002), Interaction of A2E with model membranes. Implications to the pathogenesis of age-related macular degeneration, *J. Gen. Physiol.* 120, 147-157.

-S. Bhattacharya and S. De, (1999), Synthesis and vesicle formation from dimeric pseudoglyceryl lipids with $(CH2)_m$ spacers: Pronounced *m*-Value dependence of thermal properties, vesicle fusion, and cholesterol complexation, *Chem. - A Eur. J.* 5, 2335-47.

Date submitted: 12th June 2005

Todor G. Deligeorgiev, Ph.D.

University of Sofia,
Faculty of Chemistry,
1 James Bourchier Avenue,
126 Sofia, Bulgaria.
Tel: +359 2 816 1269   Fax: +359 2 962 5438
toddel@chem.uni-sofia.bg
www.chem.uni-sofia.bg/unidyes/

Specialty Keywords: Dye Synthesis, Fluorescence, Bioapplications of Fluorescent Probes.

In the last years our research continue to be directed to the synthesis of novel nucleic acid dyes based mainly on Thiazole Orange and Oxazole Yellow chromophores as non-covalent fluorescent probes. We are also interested in development of novel luminescent europium, aluminium and zinc complexes based on different ligands.

-Todor Deligeorgiev, Aleksey Vasilev, Karl-Heinz Drexhage, Quaternization of N-Heterocycles with acrylamide and N-alkyl acrylamides, *Dyes and Pigments,* 67, 21 (2005).

-Todor Deligeorgiev, Aleksey Vasilev, Karl-Heinz Drexhage, Synthesis of novel monomeric and homodimeric cyanine dyes based on oxazolo[4,5-b]pyridinium and quinolinium end groups for nucleic acid detection, *Dyes and Pigments,* 66(2), 135-142 (2005).

Date submitted: 27th June 2005

James N. Demas, Ph.D.

University of Virginia,
Department of Chemistry,
Charlottesville, VA 22904,
USA.
Tel: 434 924 3343   Fax: 434 924 3710
demas@virginia.edu
www.people.virginia.edu/~jnd/

Specialty Keywords: Coordination Compounds, Luminescence, Sensors.

We are designing, synthesizing, and applying highly luminescent Ru, Os, Ir, and Re complexes with α-diimine ligands. Applications are as molecular reporters and analytical sensors (e.g., O_2 and pH) with special focus on the role of the support in modulating and controlling sensing properties. We are also developing instrumentation and data analysis methods.

-W. J. Bowyer, Wenying Xu, and J. N. Demas (2004). Determining oxygen diffusion coefficients in polymer films by lifetimes of luminescent complexes measured in the frequency domain, *Anal. Chem.*, **76**, 4374-4378.

Date submitted: 17th July 2005

Alexander P. Demchenko, Ph.D.

Research Institute for Genetic Engineering and Biotechnology,
TUBITAK, Gebze-Kocaeli, 41470,Turkey.
& Palladin Institute of Biochemistry,
Leontovicha 9, Kiev 01030, Ukraine.
Tel: 90 262 641 2300   Fax: 90 262 646 3929
dem@rigeb.gov.tr

Specialty Keywords: Protein and Membrane Fluorescence, Red-Edge Effects, New Fluorescence Sensors and Probes.
AIM 2003 = 26.9

Based on Red-Edge effects in fluorescence a new methodology was developed for the studies of protein and biomembrane dynamics. A new generation of two-color ratiometric fluorescence probes and sensors was developed based on 3-hydroxychromones were developed and applied for the studies of polarity, hydration and electrostatic potentials in biomembranes and living cells. Other research interests include protein folding, protein-ligand interactions and the functioning of molecular chaperones.

-A.P. Demchenko. Optimization of fluorescence response in the design of molecular biosensors (Review). Anal. Biochem. 343 (2005) 1-22.

Date submitted: 11th August 2005 **Richard B. Dennis, Ph.D.**

Gilden Photonics Ltd.,
480 Lanark Road,
Edinburgh EH14 5BL,
Scotland, UK.
Tel: +44 (0)131 442 4406
Richard.Dennis@gildenphotonics.com
www.gildenphotonics.com

Specialty Keywords: Fluorescence, Spectroscopy, Time Resolved.

Actively involved since 1978 in the development of fluorescence measurement techniques, particularly time correlated single photon counting for time resolved studies with resolution down to picoseconds and ultra-sensitive, high resolution fluorescence spectroscopy. Co-founder in 2005 of new specialist company, Gilden Photonics Ltd., a specialist manufacturing and representative company focused on the design, manufacture, sales and service of spectroscopic components and optical instruments.

Date submitted: 20th October 2006 **Richard J. DeSa, Ph.D.**

On-Line Instrument Systems, Inc.,
Olis Inc.,
130 Conway Drive, Suites A, B, & C,
Bogart, GA 30622, USA.
Tel: 001 706 353 654   Fax: 001 706 353 1972
Chief@olisweb.com
www.olisweb.com
Specialty Keywords: Rapid-scanning Emission, Rapid-scanning Excitation, Stopped-flow, Circularly Polarized Luminescence, FDCD.

Dr. Desa, is the Founder and Present On-Line Instrument Systems, Inc.
US Patent 6,970,241 B1, Nov 29, 2005. "Device for Enabling Slow and Direct Measurement of Fluorescence Polarization"
-Recording Polarization of Fluorescence Spectrometer - A Unique Application of Piezoelectric Birefringence Modulation," John E. Wampler and Richard J. DeSa, Analytical Chemistry, 46563 (1974).
-An On-Line Spectrofluorimeter System for Rapid Collection of Absolute Luminescence Spectra," John E. Wampler and Richard J. DeSa, Applied Spectroscopy, 25, No. 6, 623-627 (1971).

Devaney, J. J.
Diaspro, A.

Date submitted: 6th April 2006

John J. Devaney.

Boston Electronics Corporation,
91 Boylston Street, Brookline,
MA 02445,
USA.
Tel: 800 347 5445 or 617 566 3821   Fax: 617 731 0935
devaney@boselec.com
www.boselec.com

Specialty Keywords: TCSPC, Spectroscopy, Photodetection.

John Devaney is an instrumentation Engineer at Boston Electronics Corporation, North American agents for Becker & Hickl GmbH of Berlin, Germany and for Edinburgh Instruments Ltd of Edinburgh, Scotland. Specialist in TCSPC applications.

Date submitted: 8th June 2006

Alberto Diaspro, Ph.D.

LAMBS-MicroScoBio, Department of Physics,
University of Genoa,
Via Dodecaneso 33, Genoa,
Liguria, 16146, Italy.
Tel: +39 010 353 6426   Fax: +39 01 031 4218
diaspro@fisica.unige.it
www.lambs.it

Specialty Keywords: Confocal Microscopy, Multiphoton
Microscopy, Bioimaging, Single Molecule Imaging, Nano.

AD (Genoa, Italy, 1959) is Professor at the University of Genoa in the Applied Physics area. His research activity aims at the study of biological molecules to address cell functioning using conventional, confocal and multiphoton fluorescence microscopy, FRET, FRAP, FLIM, single molecule imaging, scanning probe microscopy, polarized light scattering, nanostructured model systems, bioimaging. He joins IFOM (Institute for Cancer Research, Milan), MicroScoBio (Research Center for Correlative Microscopy in Biomedicine and Oncology) and NANOMED National strategic program.

-DIASPRO A. (2006). Shine on ... proteins. Microsc.Res.Tech vol. 69, pp. 149-151; DIASPRO A., et al. (2005). Two-photon fluorescence excitation and related techniques in biological microscopy. Quart. Rev. Biophys. vol. 38(2), pp. 1-72.

Date submitted: 9th June 2005

Andrzej T. Dobek, Ph.D.

Faculty of Physics, A.Mickiewicz University in Poznań,
Umultowska 85,
61-614 Poznań,
Poland.
Tel: +48 61 829 5252   Fax: +48 61`825`7018
dobek@amu.edu.pl
bio5.physd.amu.edu.pl

Specialty Keywords: Molecular Biophysics, Photobiology, Ultra-fast Laser Spectroscopy.

Current Research Interests: Transient absorption, fluorescence and photovoltage studies of primary events in photosynthesis, static and dynamic light scattering in biomacromolecular solutions, nonlinear light scattering in solution of macromolecules oriented by DC magnetic field and optical field.

-A.Włodarczyk, P.Grzybkowski, A.Patkowski, A.Dobek (2005). Effect of ions on the polymorphism, effective charge, and stability of human telomeric DNA. Photon correlation spectroscopy and circular dichroism studies, *J.Phys.Chem. B*, 109, 3594-3605.

-H. Jurga-Nowak, E. Banachowicz, A. Dobek, A. Patkowski (2004). Supramolecular guanosine 5c-monophosphate structures in solution. Light scattering study, *J. Phys.Chem. B*, 108, 2744-2750.

Date submitted: 5th April 2006

Wen-Ji Dong, Ph.D.

School Chemical Engineering & Bioengineering,
Dept. VCAAP, Washington State University, Wegner 205,
Pullman, WA 99164-6520,
USA.
Tel: 509 335 5798   Fax: 509 335 4650
wdong@vetmed.wsu.edu

Specialty Keywords: Fluorescence, Lanthanide Luminescence, FRET, Kinetics, Cardiac Thin Filament Proteins, Bioassay.
AIM 2005 = 9.1

The primary focus of my current research involves the application of fluorescence spectroscopy combining with molecular biology and fast kinetics approaches in study of cardiac thin filament regulation and bioassay development, including study of calcium activation mechanism of cardiac muscle; elucidation of structure-function relationship within thin filament; and development and application of novel fluorescence and luminescence assay for biological studies, high throughput drug screening and cardiac marker detections.

-Brouillette, C. G., Dong, W. J., Yang, Z. W., Ray, M. J., Protasevich, II, Cheung, H. C., and Engler, J. A. (2005) Forster resonance energy transfer measurements are consistent with a helical bundle model for lipid-free apolipoprotein A-I. *Biochemistry 44*, 16413-25.

Doré, K.
Doroshenko, A. O.

Date submitted: 17th October 2006

Kim Doré, (Ph.D. Candidate)

Department of Chemistry and,
Centre d'optique, photonique et laser (COPL),
Université Laval,
Québec City, PQ, Canada, G1K 7P4.
Tel: 418 656 2131 Ex: 8681   Fax: 418 656 7916
kim.dore.1@ulaval.ca

Specialty Keywords: Biosensors, Superlighting, RET.
AIM 2006 = 8.2
Ultrasensitive detection of genetic material using a polymeric biosensor. Development of a method allowing the hybridization and detection of unfragmented genetic DNA in pure water. Characterization of an intrinsic fluorescence amplification mechanism based on RET in supramolecular aggregates. Based on light scattering and TEM results, a physical model having a tubular form was determined to be accurate.

-Doré K, Leclerc M, Boudreau D (2006) Investigation of a fluorescence signal amplification mechanism used for the direct molecular detection of nucleic acids. J Fluoresc 16:259-265.
-Doré K, Neagu-Plesu R, Leclerc M, Boudreau D, Ritcey A. M. (2006) Characterization of superlighting polymer-DNA aggregates: a fluorescence and light scattering study. Langmuir - in press.

Date submitted: 12th April 2006

Andrey O. Doroshenko, Ph.D., D.Sc.

Department of Organic Chemistry,
Kharkov V.N.Karazin National University,
4 Svobody Sqr., Kharkov,
61077, Ukraine.
Tel: +38 057 707 5335   Fax: +38 057 707 5130
andrey.o.doroshenko@univer.kharkov.ua
www.cesj.com/doroshenko

Specialty Keywords: High Stokes Shift Organic Luminophores.

Design and investigation of abnormally high Stokes shift organic fluorescent species: sterically hindered aromatic/heterocyclic molecules, excited state intramolecular proton transfer (ESIPT) compounds, cation-sensitive fluorescent probes, fluorescent probes for biomembrane studies. Elucidation of interrelations between the molecular structure, photophysical and photochemical properties of organic compounds. Photochemical transformations of organic molecules.
-Quantum chemical modeling of fluorescent and photochemical ability of organic luminophores.
Sizova Z.A., Doroshenko A.O., Lukatskaya L.L., Rubtsov M.I., Karasyov A.A. 2004,
-J. Photochem. Photobiol., A: Chem., 165, 59-68.
Kolos N.N., Paponov B.V., Orlov V.D., Lvovskaya M.I., Doroshenko A.O., Shishkin O.V. 2006,
J. Molec. Struct., 785, 114-122.

Date submitted: 9th August 2005 **Peter Douglas, Ph.D.**

Chemistry Department,
University of Wales Swansea,
Singlton Park, Swansea,
SA2 8PP, UK.
Tel: +44 (0) 179 225 1308 Fax: +44 (0) 179 229 5747
P.Douglas@swan.ac.uk

Specialty Keywords: Porphyrins, Optical Sensors, Photographic Dyes.
Photochemical research interests: Photodegradation mechanisms of photographic and textile dyes, photochemistry on thin film TiO_2, luminescent oxygen sensors, thin film optical sensors for medical industrial and environmental applications, photochemistry of porphyrins and metalloporphyrins, colloidal photochemistry electrochemistry and reaction kinetics.

-C.D.Geddes and P.Douglas, Fluorescent dyes bound to hydrophilic copolymers - applications for aqueous halide sensing, (2000), *App. Poly. Sci.,* 76, 603-615.
-P.Douglas and K.Eaton, Response characteristics of thin film oxygen sensors, Pt and Pd Octaethylporphyrins in polymer films, (2002) Sens. Actuators B, 200-208.

Date submitted: 5th April 2006 **Cathrin Dressler, Ph.D.**

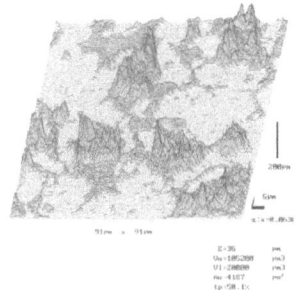

Laser- und Medizin-Technologie GmbH.,
Berlin, Fabeckstrasse 60-62,
D-14195 Berlin,
Germany.
Tel: +49 (0) 308 449 2326 Fax: +49 (0) 308 449 2399
c.dressler@lmtb.de
www.lmtb.de

Specialty Keywords: Cell Stressing, Nanocrystals, Subcellular Structures.
My research is focused on analyzing cellular stress responses by means of fluorescence microscopy and scanning probe microscopy. Especially we are interested in new organic fluorescent labels as well as luminescent nanocrytals (quantum dots). The development of nanostructered substrates in fluorescence-based bioanalytical devices also is a main working area of our group.
-Dressler C, Eberle H-G, Beuthan J, Müller G. 2002. Microscopic techniques in bioanalytics and micropeparation of cellular systems. In: *Optoelectronics applications in medicine, food technology and environmental protection*; pp. 95-103; Ecomed Verlagsges., Landsberg
-Dressler C, Minet O, Novkov V, Müller G, Beuthan J (2005). Microscopical heat stress investigations under application of quantum dots. *J. Biomed. Optics* 10: 1-9.

Drexhage, K. H.
Duportail, G.

Date submitted: 29th August 2005

Karl H. Drexhage, Ph.D.

Chemistry Department, University of Siegen,
D-57068 Siegen,
Germany.
ATTO-TEC GmbH, D-57008 Siegen, Germany.
Tel: +49 (0) 271 740 4187   Fax: +49 (0) 271 742 0502
drexhage@chemie.uni-siegen.de
drexhage@atto-tec.com

Specialty Keywords: Fluorescence, Organic Dyes,
Fluorescent Labels.

Research Interests: My research is centered around the process of light emission by molecules and the influence of molecular structure on fluorescence. Research topics are: Inter- and intramolecular energy transfer, influence of a mirror on decay time and directional characteristics of fluorescence, cooling by anti-Stokes fluorescence, laser dyes, development of fluorescent labels for biochemistry and medicine.

-J. Arden-Jacob, J. Frantzeskos, N.U. Kemnitzer, A. Zilles, and K.H. Drexhage (2001). New fluorescent markers for the red region, Spectrochim. Acta A, 57(11), 2271-2283.

Date submitted: Editor Retained.

David T. F. Dryden, Ph.D.

School of Chemistry, University of Edinburgh,
The King's Buildings, Edinburgh,
EH 9 3JJ,
United Kingdom.
Tel: +44 (0) 131 650 4735
david.dryden@ed.ac.uk
www.chem.ed.ac.uk/staff/dryden.html & www.cosmic.ed.ac.uk/

Specialty Keywords: Protein-DNA Interactions, Fluorescence
Spectroscopy, Single-Molecule Imaging.

I am interested in all aspects of protein and DNA structure and dynamics with particular emphasis on combining physical and biological techniques at the "interface" between the physical and life sciences.

-M.D. Walkinshaw, P. Taylor, S.S. Sturrock, C. Atanasiu, T. Berge, R.M. Henderson, J.M. Edwardson, and D.T.F. Dryden. Structure of Ocr from Bacteriophage T7, a Protein that Mimics B-Form DNA. *Molecular Cell* [2002] 9, 187–194.
-The DNA binding characteristics of the trimeric *Eco*KI methyltransferase and its partially assembled dimeric form determined by fluorescence polarisation and DNA footprinting. L.M. Powell, B.A. Connolly & D.T.F. Dryden. [1998] *J. Mol. Biol.* 283, 947-961.

Date submitted: 3rd July 2006

Guy Duportail, Ph.D.

Laboratoire de Pharmacologie et Physicochimie,
Equipe "Photophysique des Interactions Biomoléculaires,
C.N.R.S. & Université Louis Pasteur de Strasbourg,
67401 Illkirch, France.
Tel: (33) 39 024 4260   Fax: (33) 39 024 4312
Guy.Duportail@pharma.u-strasbg.fr

Specialty Keywords: Biophysics, Liposomes, Fluorescent Probes
AIM 2004 = 12.7

Field of research is membrane biophysics by using fluorescence spectroscopy methods. The different topics of interest so far considered are: membrane photophysics, development of liposomes as mimicking membrane systems, conception and study of novel fluorescent probes for biomembranes (DPH, Pyrene and 3-Hydroxyflavone derivatives), physicochemistry of the processes of non-viral transfection by cationic lipids, lipid microdomains (rafts).

-G. Duportail, and P. Lianos (1996) in *Vesicles,* Surfactant Science Series, Vol. 62 (M. Rosoff Ed.), Marcel Dekker, New York, pp. 295-372.

-A.S. Klymchenko, G. Duportail, A.P. Demchenko, and Y. Mély (2004) Bimodal distribution and fluorescence response of environment-sensitive probes in lipid bilayers. *Biophys. J.* 86:2929-2941.

Date submitted: 11th May 2006

Axel Dürkop, Ph.D.

Institute of Analytical Chemistry,
University of Regensburg,
Universitätsstraße 31,
93053 Regensburg, Germany.
Tel: 0049 941 943 4053   Fax: 0049 941 943 4064
axel.duerkop@chemie.uni-regensburg.de

Specialty Keywords: Ruthenium Complexes, Saccharides, DNA-Intercalators, Lanthanide Complexes, Labels, Phosphate.

Our research includes the synthesis of new luminescent probes and metal ligand complexes to be conjugated to functional groups in biomolecules for bioassays (e.g. (FRET)-Immunoassays, (FRET)-DNA-assays) or to act as DNA-Intercalators. All assays are developed in microplates. Further work is done on the fluorescent analysis of phosphate, metal ions or enzyme activity.

-A. Dürkop, M. Turel, A. Lobnik, O. S. Wolfbeis, (2006), Microtiter Plate Assay for Phosphate Using a Europium-Tetracycline Complex as a Sensitive Luminescent Probe, *Anal. Chim. Acta.,* 555, 292-298.

-Dürkop A., Schaeferling M., Wolfbeis O. S., (2006), Glucose Sensing and Glucose Determination Using Fluorescent Probes and Molecular Receptors, in *Topics in Fluorescence Spectroscopy,* Vol. 11, (Geddes C. D., Lakowicz J. R., Eds), Springer, New York, 351-376.

Dyubko, T. S.
Eastwood, D.

Date submitted: Editor Retained.

Tatyana S. Dyubko, Ph.D.

Institute for Problems of Cryobiology and Cryomedicine,
Ukrainian National Academy of Sciences,
23 Pereyaslavskaya Str., Kharkov,
61015, Ukraine.
Tel: + 380 57 772 6141   Fax: + 380 57 772 0084
cryo@online.kharkov.ua & (Inst.); dyubko@un.com.ua
www.geocities.com/tdyubko2003/index.htm

Specialty Keywords: Fluorescence, Biophysics, Cryobiology.

The main research ones is including: (1) development of fluorescent probe methods and its application to determination of human serum proteins and biological membranes structural rearrangements after non-physiological conditions action (low temperatures, laser and ionizing radiation etc.) and under some human diseases; (2) testing of new fluorescent dyes with aim of its application in biology and medical diagnostics; (3) investigation of molecular mechanisms of cell membranes cryodamages and cryoprotection. Author of more 135 scientific works.

-Dyubko T.S. Cell membrane cryodamages according to spectroscopy of fluorescent probes data. Journal of Biosciences, 1999, v. 24, suppl. 1, p. 248.

-Romodanova E.A., Dyubko T.S. et al. MNBIS as marker of protein macrostructure changes. Bulletin of KhNU, No 570. Ser. Radiophysics and Electronics, 2002.Is. 2, p. 302.

Date submitted: 26th September 2006

DeLyle Eastwood, Ph.D.

Chemistry Department,
University of Idaho,
Rayburn, Moscow, Latah,
ID 83844-2343, USA.
Tel: 208 885 7785   Fax: 208 885 6173
delyle@uidaho.edu

Specialty Keywords: Envtl Anal, Sensors, Nanotech, Porphyrins/dyes. Uranium trioxide.

Nanotechnology,-photoluminescence of semiconductor quantum dots, nanotubes, films of fullerene/phthalocyanine (Pc). Ultratrace fluorescence analysis explosives. HomelandSecurity environmental sensors/field instruments for uranyl in water/soil, PCBs, petroleum oils PAHs for oil ID(CG), marine luminescence, Pt .Pd, Zn. Cu porphyrins (P) luminescence properties.

-G. Li, L. W. Burggraf, J. R. Shoemaker, D. Eastwood, and A. E. Stiegman, *High-Temperature Photoluminescence in Sol-Gel Silica Containing SiC/C Nanostructures*, Applied Physics Letters 76, 3373 (2000).

-D. Eastwood, C. Fernandez, B.Yoon, C. N. Sheaff, and C. M. Wai. *Fluorescence of Aromatic Amines and Their Fluorescamine Derivatives for Detection of Explosive Vapors.* Appl. Spectrosc. 60(9), 958-963, 2006.

Date submitted: Editor Retained.   **Kay Eaton, Ph.D.**

Chemistry Department,
University of Wales Swansea,
Singlton Park, Swansea,
SA2 8PP, UK.
Tel: +44 (0) 179 229 5506   Fax: +44 (0) 179 229 5747
cmsolar@swan.ac.uk

Specialty Keywords: Optical Oxygen Sensors, Redox Chemistry, Luminescence Quenching.

Research interests: The development of novel luminescence and redox based optical oxygen sensors. Dye redox chemistry. Steady-state and time-resolved studies of metalloporphyrin luminescence quenching by oxygen. Kinetic modelling of oxygen quenching of luminescence in heterogeneous thin polymer films.

-K. Eaton, A novel colorimetric oxygen sensor: dye redox chemistry in a thin polymer film, (2002), Sens. and Actuators B, 85, 42-51.
-P.Douglas and K.Eaton, Response characteristics of thin film oxygen sensors, Pt and Pd Octaethylporphyrins in polymer films, (2002) Sens. Actuators B, 82, 200-208.

Date submitted: 26th September 2006   **Richard H. Ebright, Ph.D.**

Howard Hughes Medical Institute,
Rutgers University,
190 Frelinghuysen Road,
Piscataway, NJ 08902, USA.
Tel: 732 445 5179   Fax: 732 445 5735
ebright@waksman.rutgers.edu
www.hhmi.org/research/investigators/ebright.html
Specialty Keywords: Transcription, Labelling strategies, FP,
Single-molecule imaging, Single-molecule nanomanipulation.
AIM 2005 = 86.0

Our group is interested in the first step in gene expression: i.e., transcription. Our objectives are: (i) to understand the structural and mechanistic basis of transcription initiation, (ii) to understand the structural and mechanistic basis of transcription elongation, and (iii) to develop inhibitors of bacterial transcription for application as antibacterial agents.

-Kapanidis, A., Margeat, E., Laurence, T., Doose, S., Ho, S.O., Mukhopadhyay, J., Kortkhonjia, E., Mekler, V., Ebright, R., and Weiss, S. (2005) Retention of transcription initiation factor σ^{70} in transcription elongation: single-molecule analysis. *Mol. Cell* 20, 347-356.

Egelhaaf, H.-J.
Ehrenberg, B.

Date submitted: Editor Retained.

Hans-Joachim Egelhaaf, Ph.D.

Institute of Physical Chemistry, University of Tübingen,
Auf der Morgenstelle 8, D-72076 Tübingen,
Germany.
Tel: +49 7071 297 6911   Fax: +49 7071 297 5490
hans-joachim.egelhaaf@ipc.uni-tuebingen.de
uni-tuebingen.de/hans-joachim.egelhaaf/

Specialty Keywords: Thin Organic Films, Molecular Mobility, Fluorescence Anisotropy.

Translational and rotational molecular mobilities in liquid-swollen polymers are investigated by steady-state and time-resolved fluorescence techniques (mainly quenching and anisotropy) in order to understand and control the accessibilities of polymer-bound active centers.

Photoinduced processes (e.g., charge carrier generation and recombination) in thin organic films of Pi-conjugated polymers are studied by steady-state and time-resolved absorption, fluorescence, and photoconductivity in order to elucidate the kinetics and mechanisms of these processes.

-H.-J. Egelhaaf, D. Oelkrug, P. Herman, E. Holder, H.A. Mayer, E. Lindner (2001) *J. Mater. Chem.* 11, 2445 – 2552.
-G. Cerullo, G. Lanzani, S. deSilvestri, H.-J. Egelhaaf, L. Lüer, D. Oelkrug (2000) *Phys. Rev. B* 62, 2429.

Date submitted: 8th June 2006

Benjamin Ehrenberg, Ph.D.

Department of Physics,
Bar Ilan University,
Ramat Gan,
IL-52900, Israel.
Tel: 009 723 531 8427   Fax: 009 723 535 3298
ehren@mail.biu.ac.il
www.ph.biu.ac.il/fac.php?name=ehrenberg

Specialty Keywords: Fluorescent probes, Photosensitizers, Porphyrins, membrane biophysics.

We study the interactions of porphyrins and porphyrin-like molecules with artificial and natural membranes. The porphyrins are considered for use as photosensitizers for photodynamic therapy of malignancies and bacterial eradication. The aim of these studies is to understand the binding efficiency and topography of porphyrin sensitizers in membranes and to correlate these attributes with molecular structure. The extent of interaction, the depth of membrane-penetration and the efficiency of sensitized generation of singlet oxygen are monitored by fluorescence techniques.

-Ben-Dror S, Bronshtein I, Wiehe A, Röder B, Senge MO, Ehrenberg B. On the correlation between hydrophobicity, liposome binding and cellular uptake of porphyrin sensitizers. *Photochem. Photobiol.* 82, 695-701 (2006).

Date submitted: 30th June 2005   **Jörg Enderlein, Ph.D.**

Institute for Biological Information Processing 1,
Forschungszentrum Jülich,
D-52425 Jülich,
Germany.
Tel: +49 246 161 8069   Fax: +49 246 161 4216
j.enderlein@fz-juelich.de
www.joerg-enderlein.de

Specialty Keywords: Single Molecule Spectroscopy, TCSPC, Fluorescence Fluctuation Spectroscopy, Nano-Optics.

Jörg Enderlein works in the area of single-molecule spectroscopy since over ten years. His special interests are the improvement and application of new spectroscopic methods, such as fluorescence fluctuation spectroscopy, time-correlated single-photon counting, confocal microscopy, and wide-field imaging for single-molecule studies. Additional topics of research are metal-fluorophore interactions and nanooptics. He is head of the Single-Molecule Spectroscopy Group at the Forschungszentrum Jülich in Germany.

Date submitted: 29th June 2005   **Yves Engelborghs, Ph.D.**

Biomolecular Dynamics, Katholieke Universiteit Leuven,
Celestijnenlaan 200 D, Leuven,
Belgium, B3001,
Belgium.
Tel: 32 16 32 7160   Fax: 32 16 32 7974
Yves.Engelborghs@fys.kuleuven.be
www.chem.kuleuven.be/research/bio/webye_en.html

Specialty Keywords: Protein Fluorescence, Tryptophan, FCS, FCCS.
AIM 2004 = 55.5

We are very interested in protein dynamics in vitro and in the living cell. We study tryptophan (W) fluorescence in great detail and try to link fluor.lifetimes and W-rotamers. We are especially interested in unraveling the fluorescence of multiple W-proteins, and the interactions among the W's. In the context of cellular dynamics we apply FCS and FCCS to study molecular interactions in the living cell with a special interest in the proteins from HIV and nuclear factors.

-Maertens G, et al. (2005) Measuring protein-protein interactions inside living cells using single color fluorescence correlation spectroscopy. Application to human immunodeficiency virus type 1 integrase and LEDGF/p75. FASEB J. 19, 1039-41.

Date submitted: 3rd May 2005 — **Rainer Erdmann, Dipl.-Phys.**

PicoQuant GmbH.,
Rudower Chaussee 29,
12489 Berlin,
Germany.
Fax: +49 (0)306 392 6560
photonics@pq.fta-berlin.de
www.picoquant.com

Specialty Keywords: Pulsed Diode Lasers, Time-resolved Spectroscopy, Single Molecule Detection, TCSPC.

Current Status: Managing Director at PicoQuant GmbH.

We focus our R/D on ultra sensitive fluorescence detection methods. Beside the development of components (like compact picosecond diode lasers, PC boards for TCSPC, detector modules) we design complete fluorescence spectrometers for various applications including comprehensive data analysis tools. Furthermore we develop microscope based systems for fluorescence lifetime imaging (FLIM) applications. These systems offer ultimate sensitivity as well as highest spatial resolution as needed for single molecule detection. Beside traditional fluorescence correlation and fluorescence lifetime analysis we work on the combination of both techniques.

-Benda A., Hof. M., Wahl M., Patting M., Erdmann R., Kapusta P., Review of Scientific Instruments, Vol.76, 033106 (2005).

Date submitted: 18th April 2006 — **Sergei A. Eremin, D.Sc.**

M.V.Lomonosov Moscow State University,
Department of Chemical Enzymology,
Faculty of Chemistry,
Leninski Gori 1, Moscow, 119992, Russia.
Tel: +7 095 939 4192   Fax: +7 095 939 2742
eremin@enz.chem.msu.ru
www.enzyme.chem.msu.ru/eremin/

Specialty Keywords: Fluorescence Polarization Immunoassay.
AIM 2005 = 11.8

Dr. Eremin's current interests include: Development of fluorescence polarization immunoassay; Synthesis of immunoreagents and fluorescein labelled tracers and Influence of its chemical structures on the sensitivity and specificity of immunoassay.

-Sergei A. Eremin, Dietmar Knopp, Reinhard Niessner, Ji Youn Hong, Song-Ja Park, and Myung Ja Choi. High throughput determination of BTEX by one-step fluorescence polarization immunoassay. *Environ. Chem.*, 2, 227-234 (2005).
-Eremin S.A., Murtazina N.R., Ermolenko D.N., Zherdev A.V., Mart'ianov A.A., Yazynina E.V., Michura I.V., Formanovsky A.A., Dzantiev B.B. Production of polyclonal antibodies and development of fluorescence polarization immunoassay for sulfanilamide. Anal. Lett., 38: 951-969, 2005.

Date submitted: 30th June 2005

János Erostyák, Ph.D.

Department of Experimental Physics,
University of Pécs,
Ifjúság u. 6., Pécs,
H-7624, Hungary.
Tel: +36 7250 3600 / 4488   Fax: +36 7250 1571
erostyak@fizika.ttk.pte.hu
physics.ttk.pte.hu/erostyak/

Specialty Keywords: Integrating Sphere, Dielectric Relaxation, Energy Transfer, Fluorescence Tracing.

Present research interests are: Dielectric relaxation of dyes and proteins; Intra- and intermolecular energy transfer; Computer modelling of excited state processes; Analytical applications of dye-trace detection; Development of integrating spheres.
Experimental practice: phase fluorometry, femtospectrometry, laser fluorometry.

-A. Buzády, J. Savolainen, J. Erostyák, P. Myllyperkiö, B. Somogyi, J. Korppi-Tommola: Femtosecond transient absorption study of excitation relaxation of an acrylodan dye in solution and attached to human serum albumin. J. Phys. Chem. B, (2003), 107, 1208-1214.

Date submitted: 27th June 2005

Kadriye Ertekin, Ph.D.

Department of Chemistry,
Faculty of Sciences and Arts,
University of Dokuz Eylul Buca Izmir,
Turkey.
Tel: 90 232 412 8689
Kadriye.ertekin@deu.edu.tr
people.deu.edu.tr/kadriye.ertekin/

Specialty Keywords: Optical Sensor, pH Sensor, Carbon Dioxide Sensing.

I'm interested in new and bright fluorescent pH indicators, especially the ones which have pKa values between 7.4 –10.0 for CO_2 and/or HCO_3^- sensing purposes. My aim is to investigate such kind of indicator dyes in different matrices in terms of quantum yield, photo-stability, acidity constant (pKa) and to evaluate their performances as CO_2/ HCO_3^- sensing agents.

-Kadriye Ertekin, Ingo Klimant, Gerhard Neurauter and Otto S. Wolfbeis, Chracterization of a rezervoir-type capillary optical microsensor for pCO_2 measurments. *Talanta*, 59 261-267 (2003).

-K. Ertekin, S. Cinar, T. Aydemir, S. Alp, Sol-Gel based Glucose Biosensor Employing Fluorescent pH indicator; Dyes and Pigments (67) 2005, 133-138.

Date submitted: 29th August 2004

Jose Paulo S. Farinha, Ph.D.

Centro de Quimica-Fisica Molecular,
Instituto Superior Tecnico,
Av. Rovisco Pais, Lisboa, 1049-001,
Portugal.
Tel: 351 21 841 9221   Fax 351 21 846 4455
farinha@ist.utl.pt
dequim.ist.utl.pt/docentes/3296

Specialty Keywords: Energy Transfer, Polymers, Colloids.
AIM 2003 = 12

Study of polymer and colloidal systems using fluorescence techniques.   Use of excimer formation to study the dynamics of polymer chains in solution. Study of the interface structure, colloidal particles (latex, micelles, etc) and in polymer films using non-radiative energy transfer. Synthesis and dye-labeling of polymers.   Modeling of the energy transfer kinetics in dispersed colloidal particles, polymer blend films, and other structured materials.   Modeling of the diffusion in dye-labeled latex films. Static and dynamic fluorescence measurements.

-Farinha, J. P. S. et al *J. Phys. Chem. B* 1999, *103*, 2487.
-Farinha, J. P. S. et al *J. Phys. Chem.* 1996, *100*, 12552.

Date submitted: 17th April 2006

Karl-Heinz Feller, Ph.D.

Faculty of Medical Engineering,
University of Applied Sciences Jena,
Carl-Zeiss-Promenade 2,
Jena 07745, Germany.
Tel: +49 364 120 5621   Fax: +49 364 120 5622
feller@fh-jena.de
www.fh-jena.de/˜feller

Specialty Keywords: J-Aggregates, Fluorescence Sensors,
Picosecond Spectroscopy, Host-guest interaction.

Investigation of the effects of strong excitation induced disorder in J-aggregates. The developed theory allows the interpretation of the intensity-dependent spectra of J-aggregates as contributed from the combined action of exciton-exciton annihilation and subsequent dynamic disordering processes. Development of fluorescence sensors by means of host-guest interaction for the detection of flavor and fragrance.

-H. Glaeske, V. A. Malyshev, K.-H. Feller, Phys. Rev. A 65, 33821 – 33832 (2002).
-E. Gaizauskas, K.-H. Feller, Opt. Comm. 216, 217 – 224 (2003).
-K. Schönefeld, R. Ludwig, K.-H. Feller, J. of Fluorescence 16 (2006).

Date submitted: 24th July 2006

Bettina Felletschin, M.Sc.

Berthold Technologies GmbH. & Co. KG.,
Marketing, Calmbacher Str. 22,
Bad Wildbad, 75323,
Germany.
Tel: +49 7081 1770   Fax: +49 7081 177100
bettina.felletschin@berthold.com
www.berthold.com/bio

Specialty Keywords: Fluorescence Applications in Microplate Readers and Imaging.

The current responsibilities at Berthold Technologies as application specialist include all application questions besides marketing issues. The knowledge is based on the experience with fluorescence, time-resolved fluorescence, fluorescence polarization, BRET and luminescence assays in academic research and pharmaceutical drug discovery.

Date submitted: 21st August 2006

Roger W. P. Fenske, M.Sc.

Edinburgh Instruments Ltd.,
2 Bain Square,
Livingston, EH54 7DQ,
Scotland, UK.
Tel: +44 (0) 150 642 5300   Fax: +44 (0) 150 642 5320
rfenske@edinst.com
www.edinburghinstruments.com & www.edinst.com

Specialty Keywords: Fluorescence Spectrometers, Fluorescence Lifetime, Single Photon Counting, Instrumentation.

Roger Fenske has been active in the field of fluorescence since 2000. Roger joined Edinburgh Instruments in 2001 as a product engineer and is involved in the development, testing and installation of fluorescence spectrometers and flash photolysis instrumentation.

Date submitted: 27th June 2005

Maria L. Ferrer, Ph.D.

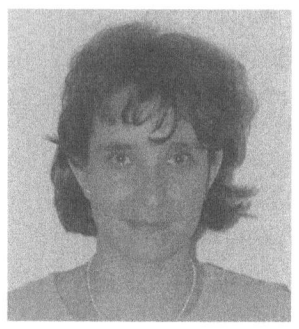

Instituto de Ciencia de Materiales,
Consejo Superior de Investigaciones Científicas,
Campus de Cantoblanco,
Madrid, 28049, Spain.
Tel: 34 91 334 9000
mferrer@icmm.csic.es

Specialty Keywords: Fluorescence Materials, Fluorescent Nanoparticles, Bioencapsulation.
AIM 2004 = 9.9

My research is focused on the preparation of organically modified silicates through the sol-gel method for optical applications. I have studied the chemical properties of the porous surface of Ormosils through fluorescence spectroscopy. More recently, I am interested on the encapsulation of biomolecules in sol-gel matrices and on the study of the structural integrity, activity and fluorescence sensing applications of the encapsulated biomolecules.

-D. Martínez-Pérez, M.L.Ferrer, C.R. Mateo (2003) A reagent less fluorescent sol-gel biosensor for uric acid detection in biological fluids.Anal. Biochem. 322 238-242.
-M.L. Ferrer, F. del Monte. (2005) Enhanced Emission in Nile Red Fluorescent Nanoparticles Embedded in Sol-Gel Hybrid Matrixes. J. Phys. Chem. B 109, 80-86.

Date submitted: 7th April 2005

Vlastimil Fidler, Ph.D.

Department of Physical Electronics,
Faculty of Nuclear Sciences and Physical Engineering,
Czech Technical University in Prague,
V Holesovickach 2, Prague 8, Czech Republic, CZ – 180 00.
Tel: +420 22 191 2221   Fax: +420 28 468 4818
fidler@troja.fjfi.cvut.cz   Vlastimil_Fidler@brown.edu
Affiliated: Brown University, Box H, Providence, RI 02912, USA.

Specialty Keywords: TR Fluorescence, Excitation Energy Transfer, Photo-induced Intramolecular Processes.

From C.V.: Graduated from Charles University in Prague, long-term stays at the Royal Institution, London, UK, at IMS Okazaki, Japan, and at University of Chicago, USA.
Current research interests: Molecular electronics; Intra-molecular electronic & vibronic energy transfer and re-distribution study, Photo-physics of molecular switching.

-V. Fidler, et al. (2002): Femtosecond Fluorescence Anisotropy Kinetics as a Signature of Ultra-fast Electronic Energy Transfer in Bichromophoric Molecules: *Z. Phys. Chem*,. 216, 589-603.
-P. Kapusta, et al. (2003): Photophysics of 3-Substituted Benzanthrones: Substituent and Solvent Control of Intersystem Crossing, *J. Phys. Chem. A.,* 107(46); 9740-9746.

Date submitted: Editor Retained.

Judit Fidy, D.Sc., Ph.D.

Dept. of Biophysics and Radiation Biology,
Semmelweis University,
Puskin u. 9, Budapest,
H-1088, Hungary.
Tel: 36 1 266 2755 / 4052   Fax: 36 1 266 6656
judit@puskin.sote.hu

Specialty Keywords: Protein Dynamics, Aggregation, Folding.

Research interests: Prof. Fidy's interest in protein dynamics started by detailed fluorescence line narrowing studies on hemoproteins. On this basis she initialized a collaboration with Prof. J. Friedrich (Bayreuth, D.) to perform the first spectral hole burning studies under high pressure on a protein. Since 1993 she has her own research lab in Budapest equipped with FLN, various luminescence methods, cryostats, high pressure cells and computer capacity for molecular modeling. They study the connection between protein dynamics and functionality.

-J. Fidy et al., invited review, BBA, (1998) 1386, 289-303.
-L. Smeller , J. Fidy, Biophys.J. (2002) 82, 426-436.

Date submitted: Editor Retained.

Jacek J. Fisz, Ph.D.

Optical Spectroscopy and Molecular Engineering Group,
Institute of Physics, N. Copernicus University,
ul. Grudziadzka 5 / 7,
PL 87-100 Torun, Poland.
Tel: +48 56 611 3297   Fax: +48 56 622 5397
jjfisz@phys.uni.torun.pl

Specialty Keywords: Molecular Fluorescence, Photochemistry, Photovoltaic Systems.

Research fields: One- and two-photon excitation spectroscopy of solutions and organized media, evanescent wave excitation fluorescence and second-harmonic generation on organized molecular assemblies, excited-state processes in solutions and organized media, structural and dynamic properties of ordered molecular media, time-resolved fluorescence spectroscopy with polarized light.

-J.J. Fisz, M.P. Budzinski, Fluorescence depolarization in organized media. Two-excited-state reactions controlled by orientation-dependent kinetic rates. I. Theory, J. Chem. Phys. 115 (15) (2001) 7130-7143.
-J.J. Fisz, A method for visual and numerical recovery of state-dependent character of fluorophore-matrix aligning interactions, Chem. Phys. Letters 355 (2002) 94-100.

Frąckowiak , D.
Fu, Y.

Date submitted: Editor Retained.

Danuta Frąckowiak, (Jabłoński) Ph.D.

Poznan University of Technology,
Institute of Physics,
Nieszawska 13A, 60 965 Poznań,
Poland.
Tel: +48 (61) 665 3180   Fax: +48 (61) 665 3201
frackow@phys.put.poznan.
www.put.poznan.pl

Specialty Keywords: Polarized Light Spectroscopy.
Current Research: 1) Investigations of the fate of absorbed energy in photosynthetic organisms, in their parts and in their anisotropic models by the measurements of polarized light fluorescence, delayed fluorescence and steady state photoacoustic spectra. The evaluation of the yield of triplet states generation using laser induced optoacoustic spectroscopy. 2)The measurements of the fluorescence of various dye-photosensitizers in healthy and cancerous cells as well as of the endogenous emission of stained cell material are due in order to select dyes suitable for photodynamic therapy and photodynamic diagnosis of cancer. From emission of irradiated stained cells the courses of photoreactions are established.

-D.J. Qian, A. Planner, J. Miyake, D. Frąckowiak (2001). Photothermal effects and fluorescence spectra of tetrapyridylporphyrins, *J. Photochem. Photobiol. A: Chemistry*, 144, 93-99.

Date submitted: 26th September 2006

Yi Fu, Ph.D.

Center for Fluorescence Spectroscopy,
Dept. of Biochemistry and Molecular Biology,
University of Maryland School of Medicine,
725 West Lombard St, Baltimore, Maryland, 21201, USA.
Tel: 410 706 7500   Fax: 410 706 8408
yifu@cfs.umbi.umd.edu
cfs.umbi.umd.edu

Specialty Keywords: Single Molecule Spectroscopy, Metal Enhanced Fluorescence, DNA.

Yi Fu's Current Interest's: Metal enhanced fluorescence using single molecule spectroscopy method.

-Fu, Y, Lakowicz JR, Enhanced Fluorescence of Cy5-Labeled DNA Tethered to Silver Island Films: Fluorescence Images and Time-Resolved Studies Using Single-Molecule Spectroscopy, Analytical Chemistry, 2006, 78, 6238-6245.

-Fu Y, Ye F, Sanders WG, Collinson MM, Higgins DA, Single Molecule Spectroscopy Studies of Diffusion in Mesoporous Silicate Thin Films, J. Physical Chemistry B, 2006, 110 (18), 9164-9170.

Date submitted: 17th August 2006

Dmitry M. Gakamsky, Ph.D.

Edinburgh Instruments Ltd.,
2 Bain Square,
Livingston, EH54 7DQ,
Scotland, UK.
Tel.: +44 (0) 150 642 5300   Fax: +44 (0) 150 642 5320
dgakamsky@edinst.com
www.edinst.com

Specialty Keywords: Fluorescence, Molecular biology, Biotechnology.

Dr. Gakamsky's current interests include: the development and application of fluorescence-based methods in molecular biology, immunology and biotechnology; fluorescence instrument design. Dr. Gakamsky has been active in the field of fluorescence since 1982. Dr. Gakamsky is the Senior Application Scientist at Edinburgh Instruments Ltd. since 2006.

Date submitted: Editor Retained.

Gerasim Stoychev Galitonov, Ph.D.

Dept of Biophysics, Inst. of Exptl Physics,
University of Warsaw,
Zwirki i Wigury 93, PL-02089 Warsaw,
Poland.
Tel: +(004822) 554 0715   Fax: +(004822) 554 0001
Gierasim@yahoo.com
www.fuw.edu.pl

Specialty Keywords: Fluorescence, Enzyme-Ligand Interactions, State-of-the-art Equipment.

Some of my interests are: Ligand tautomeric form identification and charge distribution in enzyme complexes by steady-state quenching. Rotamer identification in enzyme complexes by FRET, time-resolved and anisotropy measurements. Fluorescence and phosphorescence art-of-the-state equipment. Analysis of human genome sequences.

-Stoychev G., Kiedaszuk B. & Shugar D. (2001) Interaction of *E. coli* PNP with the cationic and zwitterionic forms of the fluorescent substrate m^{7}Guo, *BBA*, 1544 (1-2) 74-88.
-Stoychev G., Kiedaszuk B. & Shugar D. (2002) Xanthosine and xanthine: Substrate properties with PNP, and relevance to other enzyme systems, *Eur J Biochem*, 269 (16) 4048-4057.

Date submitted: Editor Retained.

Fang Gao, Ph.D.

University of Tennessee,
Department of Chemistry,
Buehler Hall 608,
Knoxville, TN 39776-1600, USA.
Tel: 865 974 3473   Fax: 865 974 3454
fgao1@utk.edu

Specialty Keywords: Dye Synthesis, Photochemistry &
Photopysics, Polymer.

Dr. Fang Gao is a research scientist at the University of Tennessee, Knoxville. His research mainly focuses on the synthesis of dyes and polymer resin, photopolymer and photochemistry. Now, he is doing the asymmetrical photochemistry. He has authored 26 journal papers. He has established his international position in these fields.

-Fang Gao, Robert N. Compton, Richard M. Pagni, The mutilphoto photochemistry of 2-iodooctane in methanol, Chemical Communications, 2003,1584-1585.
-Fang Gao, David Boyles, Rodney Sullivan, Robert N. Compton, Richard M. Pagni, The Photochemistry of racemic and resolved 2-iodooctane. The effect of solvent polarity and viscosity on the chemistry, Journal of Organic Chemistry, 2002, 67 (26), 9361-9367.

Date submitted: Editor Retained.

Michael S. Garley, Ph.D.

Chemistry Department,
University of Wales Swansea,
Singlton Park, Swansea,
SA2 8PP, UK.
Tel: +44 (0)179 229 5796   Fax: +44 (0)179 229 5747
M.S.Garley@swan.ac.uk

Specialty Keywords: Computer Modelling, Chemical Kinetics.

Research interests: Chemical kinetics, computer modelling, time resolved fluorescence and phosphorescence.

-R.J.Berry, P.Douglas, M.S.Garley, D.Clarke, C.Winscom, Triplet energies, singlet-state properties and singlet oxygen quenching rate constants and quantum yields for two cyan azamethine dyes (1999), *J.Photochem.Photobiol.A.,*120, 29-36.
-H.N.McMurray, P.Douglas, C.Busa and M.S.Garley, Oxygen quenching of tris(2,2'-bipyridine) ruthenium (II) in thin organic films, (1994) *J.Photochem.Photobiol.A.,*80, 283-288.

Date submitted: Editor Retained.

Sergiy V. Gatash, Ph.D.

Department Biological and Medicine Physics,
School of Radiophysics,
V.N. Karazin Kharkov National University,
4 Svobody Sq. Kharkov, 61077 Ukraine.
Tel: (38) 057 245 7212   Fax: (38) 057 235 3977
Sergiy.V.Gatash@univer.kharkov.ua

Specialty Keywords: Fluorescence Spectroscopy, Hydrophobic and Hydrophilic Fluorescence Probes.

Current Research Interests: My research is focused on investigation by means of fluorescence probes the conformation transitions of protein macromolecules especially fibrinogen and serum albumin.   I also study the influence of physical factors such as temperature and irradiation on conformation and function macromolecules and biological membranes.

-Gatash et al., Influence of irradiation and low temperatures on structure-dynamical state of blood proteins. // Biophysical Bulletin, Issue 2 (11), (Visn. Khark. univ.)-2002.- p.46-49.
-Andreeva et al., Influence of freezing on spectral properties of fibrinogen solutions. //Problems of Cryobiology, 1998, No 3, pp. 18-21.

Date submitted: 17th April 2006

Ehud Gazit, Ph.D.

Department of Molecular Microbiology and Biotechnology,
Tel Aviv University,
Tel Aviv 69978,
Israel.
Tel: +972 3 640 9030   Fax: +972 3 640 5448
ehudg@post.tau.ac.il
www.tau.ac.il/lifesci/departments/biotech/

Specialty Keywords: Amyloid Formation, Protein Folding, Self-Assembly, Molecular Recognition, Nanotechnology.
AIM 2005 = 65.6

In our lab we study protein folding, unfolding, misfolding and self-assembly using a variety of biochemical and biophysical techniques. A partial list of the studies systems includes amyloid fibril formation, several bacterial toxin-antidote systems, and tummro suppressor proteins. Another line of research is directed toward the study of bio-inspired nano-scale assemblies including peptide nanotubes, nanospheres, and hydrogels with nano-scale order.

-Reches, M., and Gazit, E. (2003) Casting Metal Nanowires within Discrete Self-Assembled Peptide Nanotubes. *Science* 300, 625-627.
-Cohen, T., Frydman-Marom, A., Rechter, M., & Gazit, E. (2006) Inhibition of Amyloid Fibril Formation and Cytotoxicity by Hydroxy-Indole Derivatives. *Biochemistry* 45, 4727-4735.

Date submitted: 5th July 2006

Chris D. Geddes, Ph.D.

Institute of Fluorescence,
Medical Biotechnology Center, N249,
University of Maryland Biotechnology Institute,
725 West Lombard St., Baltimore, Maryland 21201, USA.
Tel: 410 706 3149   Fax: 410 706 4600
Geddes@umbi.umd.edu
www.umbi.umd.edu/~mbc/pages/geddes.htm
Specialty Keywords: Fluorescence Sensing, Metal-Enhanced
Fluorescence, Plasmonics, Microwave-Accelerated Assays.
AIM 2005 = 102.0

Current Research Interests: All Aspects of Fluorescence, in particular: The interactions of metallic particles and surfaces with fluorophores, recently termed Metal-enhanced Fluorescence, both from a theoretical and sensing perspective. The application of low-power microwaves with surface plasmons to dramatically accelerate and optically amplify clinical and bio agent assays.

-"Metal-enhanced fluorescence-based RNA sensing" K. Aslan, J. Huang, G.M. Wilson and C.D. Geddes, J. Am. Chem. Soc, 128 (13), 4206-4207, 2006.

-"Multicolour Microwave-Triggered Metal-enhanced Chemiluminescence" K. Aslan, S. N. Malyn and C. D. Geddes, J. Am. Chem. Soc, 128, 13372-13373, 2006.

Date submitted: 29th September 2006

Marcelo H. Gehlen, Ph.D.

Departamento de Físico-Química,
Instituto de Química de São Carlos,
Av. Trabalhador São Carlense, São Carlos,
400, CEP 13566-590, Brazil.
Tel: 163 373 9952
marcelog@iqsc.usp.br
www.iqsc.usp.br
Specialty Keywords: Dyes, Charge transfer, Fluorescence
decay.
AIM 2006 = 16.8

Research interests are in the area of excited state kinetics of new dyes and molecular probes with intramolecular charge transfer. The studies are performed in solution, microheterogeneous systems (micelles and microemulsions), polymer films, glasses, and in nanometal particles using picosecond emission techniques. Applications in new optical and sensor materials are involved.

-R. V. Pereira and M. H. Gehlen (2006) Photoinduced intramolecular charge transfer in 9-aminoacridinium derivatives assisted by intramolecular H-bond *J. Phys. Chem. A* 110, 7539 – 7546.
-R. V. Pereira and M. H. Gehlen (2006) Spectroscopy of Auramine fluorescent probes free and bound to poly(methacrylic acid) *J. Phys. Chem. B* 110, 6537 – 6542.

Date submitted: 13th June 2005 · **Hans C. Gerritsen, Ph.D.**

Molecular Biophysics,
Debye Institute,
Princetonplein 1, Utrecht,
NL-3584 CC, Netherlands.
Tel: +31 30 253 2824   Fax: +31 30 253 2706
H.C.Gerritsen@phys.uu.nl
www1.phys.uu.nl/wwwmbf/

Specialty Keywords: FLIM, SPIM, CLSM, TPE.

Main areas of research are the development and application of new methodologies in fluorescence microscopy. This includes Fluorescence Lifetime Imaging, Spectral Imaging, FRET imaging, Single Molecule Imaging and Multi-Photon Excitation imaging. In addition work is carried out on the characterization of fluorescent probes and novel fluorescent markers such as quantum dots and fluorescent colloids. Applications include live cell imaging, ion concentration imaging and FRET based co-localization studies.

-Agronskaia, A. V., L. Tertoolen and H. C. Gerritsen, "Fast fluorescence lifetime imaging of calcium in living cells ", Journal of Biomedical Optics 9(6): 1230-1237, (2004).

-Photooxidation and photobleaching of single CdSe/ZnS quantum dots probed by room-temperature time-resolved spectroscopy. van Sark et al. (2001) J.Phys.Chem. B 105, 8281-8284.

Date submitted: Editor Retained. · **Ken P. Ghiggino, Ph.D.**

School of Chemistry,
University of Melbourne,
Victoria, 3010,
Australia.
Tel: +61 (3) 8344 7137   Fax: +61 (3) 9347 5180
ghiggino@unimelb.edu.au
www.chemistry.unimelb.edu.au

Specialty Keywords: Ultrafast Spectroscopy, Fluorescence Imaging, Polymer Photophysics.

Current interests: Studies of energy and electron transfer in multichromophoric assemblies using ultrafast spectroscopy techniques. Relaxation dynamics and energy migration in macromolecules studied by time-resolved fluorescence anisotropy measurements. Photophysics and time-resolved fluorescence imaging of photosensitizers for phototherapy.

-T.A. Smith, D.J.Haines and K.P. Ghiggino (2000) Steady-state and time-resolved fluorescence polarization behaviour of acenaphthene, *J. Fluorescence* 10, 365-373.

-E.K.L. Yeow, K.P. Ghiggino, J.N.H. Reek, M.J. Crossley, A.W. Bosman, P.H. Schenning and E.W. Meier (2000) The dynamics of electronic energy transfer in novel multiporphyrin functionalized dendrimers: A time-resolved fluorescence anisotropy study, *J. Phys. Chem. B* 104, 2596–2606.

Date submitted: 8th April 2006

Adam M. Gilmore, Ph.D.

Fluorescence Division,
HORIBA JobinYvon,
3880 Park Avenue,
Edison, NJ 08820-3012, USA.
Tel: 732 494 8660 Ext: 135 Fax: 732 549 5157
adam.gilmore@jobinyvon.com
www.jobinyvon.com

Specialty Keywords: Fluorescence Spectroscopy, Time-resolved Fluorescence, Global Data Analysis.

I am presently, as of January 2004, a member of the Applications team in the Fluorescence Division of Horiba Jobin-Yvon. Our team provides support for all users interested in applying high-sensitivity and high-resolution methods with Horiba JobinYvon Fluorescence instruments. I have over 10 years experience with steady-state methods and both multi-frequency and time-correlated single-photon-counting time-resolved techniques including advanced experience with global statistical data analysis.

Date submitted: 27th September 2006

Agustina Gómez-Hens, Ph.D.

Department of Analytical Chemistry, University of Córdoba,
"Marie Curie" Building Annex,
Campus of Rabanales,
Córdoba, E-14071, Spain.
Tel: 34 95 721 8645 Fax: 34 95 721 8644
qa1gohea@uco.es

Specialty Keywords: Lanthanides, Kinetic Methodology, Fluoroimmunoassays, Liposomes.
AIM 2004 = 14.1

The research interest involves the development of fluorimetric analytical methods using lanthanide ions, long wavelength fluorophores, immunoassay techniques, kinetic methodology with stopped-flow mixing technique, dry reagent technology and liposomes. The usefulness of these methods has been shown by their application in clinical, pharmaceutical, food and environmental analysis.

-A. Gómez-Hens, J.M. Fernández-Romero (2006) Analytical methods for the control of liposomal delivery systems, *Trends Anal. Chem* 25, 167-178.

-R.C. Rodríguez-Díaz, J.M. Fernández-Romero, M.P. Aguilar-Caballos, A. Gómez-Hens (2006) Chromatographic determination of flumequine in food samples by post-column derivatisation with terbium(III), *Anal. Chim. Acta.* 578, 220-226.

Date submitted: 15th June 2005

Cees Gooijer, Ph.D.

Analytical Chemistry & Applied Spectroscopy, Laser Centre,
Vrije Universiteit Amsterdam,
de Boelelaan 1083, Amsterdam, 1081 HV,
The Netherlands.
Tel: +31 20 598 7540   Fax: +31 20 598 7543
c.gooijer@few.vu.nl
www.chem.vu.nl/acas/

Specialty Keywords: Fluorescence, Phosphorescence Detection in LC and CE, Time-resolved Fluorescence, Temperature Jump.

Applied spectroscopy research is conducted in the Laser Centre Vrije Universiteit along an analytical chemistry line – in close cooperation with chromatographers with emphasis on hyphenated techniques – and a physical chemistry line focusing on the dynamics of the interaction between small molecules and (bio)macromolecules. Research topics are hyphenation of Raman spectroscopy and LC/CE; phosphorescence detection in CE; laser fluorescence detection including FRET; temperature jump/time-resolved fluorescence and cryogenic high-resolution molecular fluorescence.

-Bader, A.N., Grubor, N.M., Ariese, F., Gooijer, C., Jankowiak, R. & Small, G.J. (2004). Probing the interaction of Benzo[a]pyrene adducts and metabolites with monoclonal antibodies using fluorescence line-narrowing spectroscopy. *Analytical Chemistry, 76,* 761-766.

Date submitted: 12th Apr 2006

Karl Otto Greulich, Ph.D.

Fritz Lipmann Institute,
Beutenbergstr. 11,
Jena, D 07745,
Germany.
Tel: 49 364 165 6400   Fax: 49 364 165 6410
kog@fli-leibniz.de
www.fli-leibniz.de/greulich
Specialty Keywords: Optical Tweezers, DNA Repair, Heart, Single molecules.
AIM 2005 = 15.2

Fluorescence is used to study single DNA molecules[1] and a cell's molecular reaction to external stimuli such as UV A irradiation and optical micromanipulation. UV A induces DNA double strand breaks. Two repair mechanisms, NHEJ and HRR compete to repair such damages[2]. Micromanipulation by optical tweezers induces calcium waves, not only in excitable cardiomyocytes but also in putatively non excitable fibroblasts[3]. This may be used to develop lead substances for reducing heart damage after infarction.

- K.O.Greulich 2005 ChemPhysChem 2458 -2471;
-A.Rapp and K.O. Greulich 2004 J.Cell Sci. 117 (21), 4935 – 4945. 3) B.Perner, S.Monajembashi, A.Rapp, L.Wollweber, K.O.Greulich, 2004 Proceedings of SPIE 5514, 179 – 188.   See also: Bioforum 09/2005 40 -41

Date submitted: 18th October 2006

Ulrich-W. Grummt, Ph.D.

Friedrich-Schiller-Universitaet Jena,
Institute of Physical Chemistry,
Helmholtzweg 4, Jena,
Germany, D 07743.
Tel: +49 364 194 8350   Fax: +49 364 194 8302
cug@uni-jena.de
www.uni-jena.de/chemie/institute/pc/grummt

Specialty Keywords: Time Correlated Single Photon Counting with ps and ns Time Resolution, Polymer Photophysics.
Main research topic is photophysical chemistry of conjugated, luminescent polymers and functionalized dyes with potential application in solar energy conversion, molecular electronics, non-linear optics, optical information recording, and chemical sensing. Energy migration and electron transport are of particular interest.
Ab-initio and DFT quantum chemical calculations are used to support and interpret experimental results.
Dr. Grummt, shall be retiring in March 2006.
-E. Birckner, U.-W. Grummt, A. H. Göller, T. Pautzsch, D. A. M. Egbe, M. Al-Higari, and E. Klemm, J. Phys. Chem. A 105 (2001) 10307 – 10315.

Date submitted: 29th June 2005

Christine A. Grygon, Ph.D.

Boehringer Ingelheim Pharmaceuticals Inc.,
900 Ridgebury Road, PO Box 368,
Ridgefield, CT 06877,
USA.
Tel: 203 798 5651   Fax: 203 837 5651
cgrygon@rdg.boehringer-ingelheim.com
www.us.boehringer-ingelheim.com

Specialty Keywords: Biophysics, Ligand Binding and Kinetics, Fluorescence, Anisotropy, Imaging.
At this time, I am no longer working solely in the field of fluorescence spectroscopy as I have assumed new responsibilities as Executive Director of our biologics research group at my company. My group still makes contributions to the fluorescence field, however, and the specific responsibilities for these activities now reside with one of my senior staff, Dr. Martha Brown.

-A.L. Burd, R.H. Ingraham, S.E. Goldrick, R.R. Kroe, J.J. Crute, and C.A. Grygon (2004). Assembly of Major Histocompatability (MHC) Class II Transcription Factors: Self Association and Promoter Recognition of RFX Proteins, *Biochemistry*, 40, 12750-12760.

Date submitted: 19th July 2005

Oleksiy V. Grygorovych, Ph.D.

Department of Physical Organic Chemistry,
Institute for Chemistry,
Kharkov V. N. Karazin National University,
4 Svobody sqr., Kharkov 61077, Ukraine.
Tel: +38 05 707 5335   Fax: +38 05 707 5130
Alexey.V.Grigorovich@univer.kharkov.ua
www-chemistry.univer.kharkov.ua/dx/nii
Specialty Keywords: Complex Formation of Organic
Luminophores, Fluorescent Probes.
AIM 2004 = 1.0

Current Research Interests: Absorption and fluorescence spectroscopy of conjugated aromatic and heterocyclic organic compounds. Protolytic interactions and complexation with metal ions of conjugated aromatic and heterocyclic organic compounds in their ground and excited states. Photochemical activity of unsaturated organic compounds. Design and application of organic luminophores as new fluorescent probes and sensors for biological systems.

-Pivovarenko V. G., Grygorovych A. V., Valuk V. F., Doroshenko A. O., 2003, Journal of Fluorescence, 13, 479-487.
Munoz A., Roshal A. D., Richelme S., Leroy E., Claparols C., Grigorovich A. V., Pivovarenko V. G., 2004, Russ. J. of General Chem., 74, 438-445.

Date submitted: 31st July 2004

Yuriy A. Gryzunov, Ph.D.

Russian State Medical University,
Malaya Pirogovskaya, 1-A Moscow,
119992,
Russia.
Tel: + 7 (095) 246 4352   Fax: + 7 (095) 246 4512
gryzunov@hotbox.ru

Specialty Keywords: Proteins, Spectroscopy, Molecular
Pathology.

Steady-state and time-resolved spectroscopy, new fluorescent probes are used to study proteins and lipid-protein complexes both under physiological as well as pathological conditions. The analysis of early changes of conformation and physical-chemical properties of biomacromolecules make it possible to evaluate the state of human body in diseases and to evaluate new diagnostic and prognostic tests.

-Gryzunov YA, Arroyo A, Vigne JL, Zhao Q, Tyurin VA, Hubel CA, R E Gandley, Vladimirov YuA, Taylor RN, Kagan VE. (2003) Binding of fatty acids facilitates oxidation of cysteine-34 and converts copper-albumin complexes from antioxidants to prooxidants. *Arch Biochem Biophys* 413(1), 53-66.
-Koplik, EV, Gryzunov YA, Dobretsov GE (2003) Blood albumin in the mechanisms of individual resistance of rats to emotional stress *Neurosci Behav Physiol* 33(8), 827-832.

Gussakovsky, E. E.
Gutiérrez-Merino, C.

Date submitted: 29th September 2006   **Eugene E. Gussakovsky, Ph.D., D.Sc.**

Institute for Biodiagnostics, National Research Council Canada, Winnipeg, Manitoba R3B 1Y6 and
Department of Botany, University of Manitoba, Winnipeg, Manitoba R3T 2N2, Canada.
Tel: 204 984 4501 / 204 453 1482 / 204 223 5845
gussak@mts.net

Specialty Keywords: Fluorescence, Optical Imaging, Photobiology, Protein and Plant Sciences, Biomedicine.
AIM 2005 = 14.3

Circularly polarized luminescence, circular dichroism and other types of UV-VIS spectroscopy of biomolecules including intact and fluorescent-labeled proteins, chlorophyll-protein complexes (LHCII) in thylakoids, arteries and other biomedical tissues etc. Chiral macroaggregates of LHCII in photosynthetic apparatus. Secondary structure of proteins. Thermodynamics of protein folding/unfolding. Agricultural photobiology, solar irradiation spectra manipulation, plant growth microclimate. Biomedical application of UV-VIS spectroscopy and imaging
Dr. Michael Sowa, Institute for Biodiagnostics, NRCC, Winnipeg, Canada.
Prof. Dr. Elisha Haas, Faculty of Life Science, Bar Ilan University, Ramat Gan, Israel.
Prof. Dr. Yosepha Shahak, Institute of Horticulture, The Volcani Center, Bet Dagan, Israel.

Date submitted: 3rd October 2006   **Carlos Gutiérrez-Merino, Ph.D.**

Department Biochemistry and Molecular Biology,
Faculty of Sciences, University of Extremadura,
Avda. Elvas, s/n, 06071-Badajoz,
Spain.
Tel: / Fax: + 34 92 428 9419
carlosgm@unex.es

Specialty Keywords: FRET, Bioenergetics, Oxidative Stress.
AIM 2005 = 12.6

My research is focused on oxidative stress impairment of cellular bioenergetics and antioxidants in cell defense protection against oxidative damage. Fluorescence methodologies are currently used in my laboratory both with purified subcellular components (proteins and membranes) and with cells in culture (using digital imaging fluorescence microscopy) to measure (1) reactive oxygen species production, (2) intracellular free Ca^{2+} concentration and pH, (3) membrane potential, (4) protein-ligand interactions, and (5) FRET.
-Gutiérrez-Martín, Y., Martín-Romero, F.J., Henao, F. and Gutiérrez-Merino, C. (2005). *J. Neurochem.* 92, 973-989.
-Tiago, T., Simao, S., Aureliano, M., Martín-Romero, F.J. and Gutiérrez-Merino, C. (2006) *Biochemistry* 45, 3794-3804.

Date submitted: 13th October 2006

David J. Haines.

Cary 50 / 100 / 300 / Eclipse Team Leader,
Varian Australia Pty. Ltd.,
679 Springvale Road,
Mulgrave, 3170, Australia.
Tel: +61 39 566 1125 Fax: +61 39 566 1196
david.haines@varianinc.com
www.varianinc.com

Specialty Keywords: Steady-state fluorescence, Time-resolved fluorescence, Polarization measurements.

My background is in steady-state and time-resolved fluorescence measurements of synthetic aromatic polymers and model compounds. My current interests are the development of instrumentation for fluorescence spectroscopy, education and training in fluorescence methods, and the application of fluorescence polarization measurements. My role at Varian involves supporting fluorescence and UV-Vis customers directly, as well as Varian's global network of molecular spectroscopy specialists.

-T.A. Smith, D.J. Haines, and K.P. Ghiggino (2000). Steady-state and time-resolved fluorescence polarization behaviour of acenaphthene: *J. Fluoresc., 10, 365-373.*

Date submitted: 27th September 2006

Harri H. O. Hakala, Ph.D.

R&D, PerkinElmer Life and Analytical Sciences,
P.O. Box 10, Turku,
FIN-20101,
Finland.
Tel: +35 82 267 8693
harri.hakala@perkinelmer.com

Specialty Keywords: Lanthanide (III) Chelates.
AIM 2004 = 7.7

The main interest is to develop new, stable, luminescent lanthanide(III) chelates suitable for biochemical assays. This includes the synthesis of Eu(III), Tb(III), Sm(III) and Dy(III) chelates, study of their photophysical properties and their coupling to biomolecules either in solution or on solid-phase, as well as their use in biochemical assays.

-L. Jaakkola, J. Peuralahti, H. Hakala, V.-M. Mukkala, P. Hurskainen and J. Hovinen (2006) *J. Pept. Sci.* 12, 199-205.

-H. Hakala, V.-M. Mukkala, T. Sutela and J. Hovinen (2006), *Org. Biomol. Chem.* 4, 1383-1386.

Date submitted: Editor Retained.

Einar L. P. Hallberg, Ph.D.

Natural Sciences, Södertörns Högskola,
Alfred Nobel's Allé 7, Huddinge,
141 89,
Sweden.
Tel: +46 (8)608 4733   Fax: +46 (8)608 4510
Einar.hallberg@sh.se
www.sh.se/natur/einar.htm

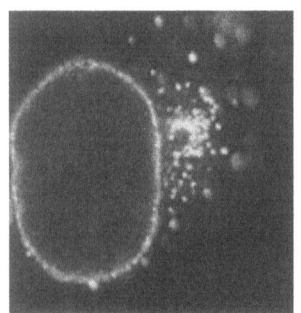

Specialty Keywords: Nuclear Membrane, Nuclear Pores, GFP.

Trafficking of proteins and RNA molecules in and out of the cell nucleus takes place via the nuclear pore complexes situated in the thousands of pores covering the nuclear surface. We investigate structural and functional aspects of how the nuclear pore complex and the nuclear membranes are organized. We use fluorescence microscopy and confocal laser scanning microscopy on cells expressing proteins tagged with GFP (Green Fluorescent Protein). We also perform Live Cell Imaging including studies of intracellular dynamics using photobleaching.

-Kihlmark, M., Imreh, G. and Hallberg, E. (2001) J. Cell Sci., 114, 3643-3653.

-Imreh, G. and Hallberg, E. (2000) Exptl. Cell Res., 259, 180-190.

Date submitted: 29th July 2004

Monika Hamers-Schneider, Ph.D.

ATTO-TEC GmbH.,
Am Eichenhang 50,
D-57076 Siegen,
Germany.
Tel: +49 (0) 271 740 4735
monika.hamers@gmx.de / hamers@atto-tec.com
www.atto-tec.com

Specialty Keywords: Fluorescent Dyes, Fluorescent Labels, Fluorescent Sensors.

My research interest is focused on the synthesis of fluorescent labels for bioanalytical applications. Furthermore I am interested in fluorescent dyes which are specially functionalized to meet the requirements of optical sensors.

-J. Arden-Jacob, J. Frantzeskos, N.U. Kemnitzer, A. Zilles, and K.H. Drexhage (2001). New fluorescent markers for the red region *Spectrochim. Acta A* 57(11), 2271-2283.
-M. Hamers-Schneider (1997). Ph.D. Thesis. Funktionelle Rhodamin-Derivate zur Fluoreszenz-Detektion in Analytik und Sensorik. Shaker Verlag, Aachen.

Date submitted: 4th April 2006

Gregory S. Harms, Ph.D.

Rudolf-Virchow-Center,
University of Würzburg,
Versbacher Str. 9, Würzburg,
D-97078 Germany.
Tel: +49 93 12 014 8717   Fax: +49 93 12 014 8718
Gregory.Harms@virchow.uni-wuerzburg.de
www.rudolf-virchow-zentrum.de

Specialty Keywords: Single-Molecule Microscopy, Dynamic Fluorescence Microscopy, Multiphoton Microscopy.
AIM 2005 = 16.3

Our goal is it to observe single bio-molecules in living systems, perform intracellular analysis, and concomitantly to develop better pharmaceutical products. We want to improve techniques of single molecule detection continuously to increase accuracy and extend the scope of application. We perform single molcule studies on of individually labelled Ion Channels (combined with patch clamping), on cytokines and cytokine receptors, and on G-Protein Coupled Receptors by wide-field imaging, by confocal techniques (FLIM and FCS), and in the near future intra-vitally with multiphoton microscopy.

Date submitted: 18th October 2006

Mary E. Hawkins, M.Sc.

National Cancer Institute, National Institues of Health,
10 Center Drive, CRC 1-3872,
Bethesda, Maryland 20892,
USA.
Tel: 301 451 7021   Fax: 301 480 1586
mh100x@nih.gov

Specialty Keywords: Fluorescent Pteridine Nucleoside Analogs.

The pteridine nucleoside analogs are very bright (quantum yields 0.4-0.9) and stable (Fidelity Systems, Inc., Gaithersburg, MD).   Native-like linkage positions probes in base-stacked orientation making fluorescence properties exquisitely sensitive to structural changes occurring nearby. This makes them extremely useful for study of structure of DNA as it binds and reacts with other molecules. We are currently working on the RNA version of the probes.
-Myers, J.C., Moore, S.A., and Shamoo, Y. (2003) *Journal of Biological Chemistry* 278 (43) 42300-42306.
-Hawkins, Mary E. Vol 7 of Topics in Fluorescence Spectroscopy, DNA Technology Chapter 5/ 151-175 (2003) Kluwer Academic/Plenum Publishers, New York Ed. Joseph R. Lakowicz.

Date submitted: 10th June 2005

Michael D. Heagy, Ph.D.

Chemistry, New Mexico Tech,
801 Leroy Ave, Socorro,
Socorro, 87801,
USA.
Tel: 505 835 5417   Fax: 505 835 5364
Mheagy@nmt.edu
infohost.nmt.edu/~chem/heagy/homepage.html

Specialty Keywords: Fluorescent Probes, Dual Fluorescence.

Research interests include the synthesis of new fluorescent systems with particular emphasis on probes for neutral biomolecules. New dual fluorescent species based on arenedicarboximide platforms have been found to display two well-resolved emission bands for ratiometric detection of specific analytes. These compounds are also being developed as flourescent proteomimetics.

-Cao, H.; McGill, T.; Heagy, M.D. "Substituent Effects on Monoboronic Acid Sensors for Saccharides Based on N-Phenyl-1,8-naphthalenedicarboximides" J. Org. Chem. 2004, 69, 2959-2966.
-Cao, H.; Heagy, M.D. "Fluorescent Chemosensors for Carbohydrates: A Decade's Worth of Bright Spies for Saccharides" J. Fluorescence 2004, 14, 567-582.

Date submitted: 15th April 2006

Ahmed A. Heikal, Ph.D.

Department of Bioengineering,
Pennsylvania State University,
231 Hallowell Building,
University Park, PA 16802, USA.
Tel: 814 865 8093   Fax: 814 863 0490
aah12@psu.edu
www.bioe.psu.edu/labs/Heikal-Lab/index.html

Specialty Keywords: Energy Metabolism, Proteins Dynamics, Biomembranes.

My laboratory focuses on understanding complex biological processes/systems on a molecular-level. We are particularly interested in understanding energy metabolism and apoptosis using NADH autofluorescence, molecular biosensors for cancer diagnosis, protein-protein interaction, and structure-function relatioship in biomembranes. Our experimental approach is a noninvasive, multimodal fluorescence micro-spectroscopy techniques with high spatio-temporal resolution.

-Vishwasrao, H.D., A.A. Heikal, K.A. Kasischke, W.W. Webb. 2005. *J. Biol. Chem.*, 208(26): 25119- 252126.
-Hess, S.T., E.D. Sheets, A. Wagenknecht-Wiesner, A.A. Heikal (2003). *Biophys. J.* 85(4), 2566-2580.

Date submitted: 5th April 2006

Ralf Heilker, Ph.D.

Boehringer Ingelheim Pharma,
Dep. of Lead Discovery,
Birkendorferstrasse 65,
Biberach an der Riss, D-88397, Germany.
Tel: +49 735 154 5590
Ralf.Heilker@bc.boehringer-ingelheim.com

Specialty Keywords: Confocal Fluorescence, High Content Screening, G-protein Coupled Receptors.

Ralf Heilker joined Boehringer Ingelheim in 1999. He was previously employed by Novartis in Basel, Switzerland (1997-1999). Throughout his career in the pharmaceutical industry he has been involved in biochemical and cellular assay development for drug discovery. In his current position as a Principal Scientist in assay development, high-throughput screening and secondary screening he is heading a research group in the Lead Discovery Department of Boehringer Ingelheim Pharma in Biberach. His particular responsibilities include the management of a High Content Screening team and the steering of GPCR target class platform. In a co-operation with the University of Ulm, he investigates novel means of GPCR-targeted drug screening using confocal fluorescence microscopy and the application of novel fluorescent proteins to High Content Screening.

Date submitted: 5th April 2006

Stefan W. Hell, Ph.D.

Department of NanoBiophotonics,
Max Planck Institute for Biophysical Chemistry,
Am Fassberg 11, 37077 Göttingen,
Germany.
Tel: +49 (0) 551 201 2500   Fax: +49 (0) 551 201 2505
shell@gwdg.de
www.mpibpc.mpg.de/groups/hell/

Specialty Keywords: Sub-Abbe Resolution, PSF- Engineering, STED, 4Pi, Saturation.

AIM 2005 = 41.8

We have introduced and developed concepts that have broken the diffraction barrier in focusing fluorescence microscopy and have attained spatial resolution at the nanometer scale. We apply these concepts, such as 4Pi and STED-microscopy, to the fluorescence imaging of live cells.

-K. I. Willig, S. O. Rizzoli, V. Westphal, R. Jahn, S. W. Hell (2006) *Nature* 440 (7086): 935- 939.
-V. Westphal and S. W. Hell (2005) *Phys. Rev. Lett.* 94: 143903.
-L. Kastrup, H. Blom, C. Eggeling, S. W. Hell (2005) *Phys. Rev. Lett.* 94: 178104.
-M. Hofmann, C. Eggeling, S. Jakobs, S. W. Hell (2005) *Proc. Natl. Acad. Sci. USA* 102 (49): 17565 – 17569.

Date submitted: Editor Retained.

Sherry L. Hemmingsen, Ph.D.

VARIAN

Varian Inc.,
Fluorescence Business Development Manager,
2700 Mitchell Drive,
Walnut Creek, CA 94598, USA.
Tel: (614) 761 1330 Fax: (614) 336 0295
sherry.hemmingsen@varianinc.com
www.varianinc.com

Specialty Keywords: Total Luminescence Spectroscopy, Fluorescence Lifetime Analysis, Instrumentation.

I support a diverse range of customer applications/needs in the life sciences, pharma, photonics, etc., develop/present fluorescence training, and contribute to marketing and sales efforts along with the development of new instrumentation and software.

Former research included fluorescence spectral and lifetime characterization of complex systems such as humic substances, chemometric methods of data analysis, Globals, MEM and total lifetime distribution analysis.

-S. L. Hemmingsen and L. B. McGown (1997). Phase-Resolved Fluorescence Spectral and Lifetime Characterization of Commercial Humic Substances: *Appl. Spectrosc.*, 57, 921.

-L. B. McGown, S. L. Hemmingsen, J. M. Shaver, L. Geng (1995). Total Lifetime Distribution Analysis for Fluorescence Fingerprinting and Characterization: *Appl. Spectrosc.*, 49, 60.

Date submitted: 5th April 2006

Manfred H. Hennecke, Ph.D.

Federal Institute for Materials Research and Testing,
Unter den Eichen 87,
Berlin, D - 12200,
Germany.
Tel: +49 308 104 1000 Fax: +49 308 104 1007
hennecke@bam.de
www.bam.de

Specialty Keywords: Chemiluminescence, Fluorescence Polarization.

Physical chemistry of polymers, in particular optical spectroscopy of dimers, oligomers and polymers (especially with polarized light, including time-resolved spectroscopy). Photochemical reactions and aging of polymers (by means of chemiluminescence).

-B. Schartel, M. Hennecke, "Thermo-oxidative stability of a conjugated polymer by chemiluminescence", Polym. Degr. Stab. 67, 249-253, 2000.

-B. Schartel, S. Krüger, V. Wachtendorf, M. Hennecke, "Excitation energy transfer of a bichromophoric cross-shaped molecule investigated by polarized fluorescence spectroscopy" J. Chem. Phys. 112, 9822-9827, 2000.

Date submitted: 27th June 2005   **Manfred Hennecke, Ph.D.**

Berthold Technologies GmbH. & Co. KG., Marketing,
Calmbacher Str. 22,
Bad Wildbad,
75323 Germany.
Tel: +49 7081 1770   Fax: +49 7081 17 7500
manfred.hennecke@berthold.com
www.berthold.com/bio

Specialty Keywords: Imaging Instruments, In-vivo
Luminescence Imaging, In-vivo NIR Fluorescence Imaging,
Gel and Blot Documentation.

The current responsibilities at BERTHOLD TECHNOLOGIES include application und instrument handling support besides marketing issues. The knowledge is based on many years of experience in imaging, in-vivo fluorescence and luminescence imaging in academic research and pharmaceutical drug discovery.

Based on this knowledge in combination with looking for the needs of researchers NightOWL II is developed with a lot of the new features. For example, controlling the excitation light beam for constant energy, quantitative fluorescence measurements will be possible.

Date submitted: 23rd July 2006   **Albin Hermetter, Ph.D.**

Institute of Biochemistry,
Graz University of Technology,
Petersgasse 12/2,
A-8010 Graz, Austria.
Tel: +43 316 873 6457   Fax: +43 316 873 6952
albin.hermetter@tugraz.at
www.biochemistry.tugraz.at/

Specialty Keywords: Lipids, Lipases, Functional Proteomics.
AIM 2004 = 33.6

Our research deals with the role of lipid hydrolysis and lipid oxidation in cellular (patho)biochemistry. In this context, we develop and apply fluorescence techniques to study lipid oxidation, the effects of oxidized lipids on intracellular signalling, and the function of lipolytic enzymes in animal and human cells on the proteome level.

-Birner-Gruenberger et al. & Hermetter, A. The lipolytic proteome of mouse adipose tissue. Mol.Cell.Proteomics 4, 1710-1717. 2005.
-Schmidinger, H., Susani-Etzerodt, H., Birner-Gruenberger, R. & Hermetter, A. Inhibitor and protein microarrays for activity-based recognition of lipolytic enzymes. Chembiochem.7, 527-534. 2006.

Hernandez-Borrell, J.
Herrmann, A.

Date submitted: 7th April 2006

Jordi Hernandez-Borrell, Ph.D.

Departament de Fisicoquímica,
Universitat de Barcelona,
Av.Joan XXIII, s.n., 08028-Barcelona,
Spain.
Tel: +34 403 5986   Fax: +34 403 5987
jordihernandezborrell@ub.edu
www.qf.ub.es/a2/nano/crew8.htm

Specialty Keywords: Biomembranes, Fluorescence, AFM.
AIM 2005 = 17.9

The lactose permease (LacY) of *E. coli* is a polytopic membrane protein often taken as a paradigm for the secondary transport. We are using LacY reconsituted into proteoliposomes and supported planar bilayers to explore the effect of the phospholipids environment in protein insertion and immobilization onto surfaces. Steady state fluorimetry, including anisotropy, surface potential methods, are used to explore the nature of the lipids present at the phospholipid-protein boundary.

-S. Merino, O. Domènech, M.T. Montero, J.Hernández-Borrell, (2005) *Langmuir* 21, 4642-4647.
-S.Merino-Montero, M.T.Montero, J.Hernández-Borrell, (2006) *Biophys. Chem.* 119, 101-105.

Date submitted: 14th June 2005

Andreas Herrmann, Ph.D.

Institute of Biology,
Humboldt-University Berlin,
Invalidenstr. 42, Berlin,
D-10115, Germany.
Tel: +49 202 093 8860   Fax: +49 202 093 8585
andreas.herrmann@rz.hu-berlin.de
www.biologie.hu-berlin.de/~molbp/new/

Specialty Keywords: Membrane, Fusion, Flip-flop.

The research is interdisciplinary ranging from structural biology, molecular and cell biology, virology, spectroscopy to the design and synthesis of biomolecule analogues. The main topics are (i) Protein-mediated fusion of membranes and (ii) Lipid-trafficking in eukaryotic cells. Essentially, Electron Spin Resonance and Fluorescence Spectroscopy/Microscopy incl. FRET, FLIM are employed. Although the main emphasise is on experimental research, the group has a very strong attraction to collaboration in theoretical modelling and simulation of experimental systems.

Date submitted: 5th April 2006

Joseph D. Hewitt, Ph.D.

Varian Analytical Instruments,
2700 Mitchell Drive,
Walnut Creek, CA 94598,
USA.
Tel: 1 800 926 3000 ext: 3064   Fax: 925 945 2360
Joe.Hewitt@varianinc.com
www.varianinc.com

Specialty Keywords: Fluorescence Instrumentation.

Dr. Hewitt's current research interests: As a sales specialist for Varian in the Southeast US, I work on application questions, Eclipse spectrofluorometer demonstrations and sales support. My individual research interests include humic substance lifetime spectroscopy, coupled detection schemes and fluorescence sensing technology.

Date submitted: Editor Retained.

Andrew R. Hind, Ph.D.

UV-Vis-NIR Sales Support Manager Europe,
Varian Ltd., 28 Manor Road, Walton-on-Thames,
Surrey, KT 12 2Qf,
England.
Tel: + 44 (0) 193 289 8000   Fax: +44 (0) 193 222 8769
andrew.hind@varianinc.com
www.varianinc.com

Specialty Keywords: Materials Science, Industrial Chemistry,
Optics / Photonics, Molecular Spectroscopy.

Background in 'applied' molecular spectroscopy research, with focus on applications in the materials science, industrial chemistry and optics/photonics areas. Experienced in the use of fluorescence, UV-Vis (including far-UV), infrared (near-, mid-, and far-) and Raman spectroscopies, with particular areas of interest including semiconductor, telecommunications, mineralogical and coating/surface characterization applications. Very interested in new spectroscopic instrumentation, techniques and applications.

-A.R. Hind, S.K. Bhargava, and S.C. Grocott (1999) Colloids Surf. A. 146, 359-374.
-A.R. Hind, S.K. Bhargava, and A. McKinnon (2001) Adv. Colloid Interfac. Sci. 93, 91-114.

Hirsch, R. E.
Hof, M.

Date submitted: 15th July 2005

Rhoda Elison Hirsch, Ph.D.

Department of Medicine (Hematology),
Department of Anatomy & Structural Biology,
Albert Einstein College of Medicine, 1300 Morris Park Avenue,
Bronx, NY 10579, USA.
Tel: 718 430 3604 Fax: 718 824 3153
rhirsch@aecom.yu.edu

Specialty Keywords: Hemoglobin, Front-face Fluorometry, Hemoglobin C Crystal Growth, Sickle Cell Hb, HbE.

Our focus is Hb mutants that give rise to disease: the ß6 hemoglobin mutants form aggregates in the red blood cell and unstable Hemoglobin E (ß26Glu→Lys). We pursue questions such as: Why does oxy HbC (ß6 Glu→Lys) form crystals in the red blood cell in contrast to deoxy sickle cell hemoglobin [HbS, ß6 Glu → Val]that forms polymers? Does HbE instability lead to disease? Additonal interests include hemoglobin based blood substitues and Hb stabilization mechanisms. Front-face fluorescence to study hemoglobin and heme-proteins is ongoing.

-RE Hirsch, "Heme Protein Fluorescence". Chapter 10 (pp. 221-255) Topics in Fluorescence Spectroscopy, Volume 6, Protein Fluorescence (ed. JR Lakowicz), New York (2000); QY Chen, -I Lalezari, RL Nagel, & RE Hirsch. "Liganded Hemoglobin Structural Perturbations by the Allosteric Effector L35." Biophysical J. 88:2057-2067 (2005).

Date submitted: Editor Retained.

Martin Hof, Ph.D.

Center for Complex Molecular Systems and Biomolecules,
J. Heyrovský Institute of Physical Chemistry,
Academy of Sciences of the Czech Republic,
Dolejškova 3, Cz-18223 Prague 8, Czech Republic.
Tel: +420 266 053 264 Fax: +420 286 582 307
Hof@jh-inst.cas.cz
www-troja.fjfi.cvut.cz/k412/home/hof/en/index.html

Specialty Keywords: Solvent Relaxation, Tryptophan Fluorescence, Fluorescence Correlation Spectroscopy (FCS).

Following main topics are presently pursued in M. Hof's laboratory:

1) Solvent relaxation in phospholipid bilayers[1]: Basic principles, applications, and new membrane labels.
2) FCS as a tool for the characterization of DNA condensation[2].
3) Formation of phospholipid mono- and bilayers controlled by FCS.
4) Picosecond tryptophan fluorescence of blood coagulation proteins.

- J. Sýkora, P. Kapusta, V. Fidler, M. Hof On What Time-Scale Does Solvent Relaxation in Phospholipid Bilayers Happen? (2002), Langmuir, 18(3), 571-574.

- T. Kral, M. Hof, M. Langner Effect of Spermine on the Plasmide Condensation and Dye Release Observed by FCS (2002), Biol. Chem. 383 (2), 331-335.

Date submitted: Editor Retained.

Johannes W. Hofstraat, Ph.D.

Dept. of Polymers & Organic Chemistry, Philips Research,
Prof. Holstlaan 4, 5656 AA Eindhoven,
Institute of Molecular Chemistry,
University of Amsterdam,
The Netherlands.
Tel: +31 40 274 4910   Fax: +31 40 274 3350
hans.hofstraat@philips.com

Specialty Keywords: Materials, Displays, Diagnostics, Photonics.

Research topics: (Electro) luminescent polymers, dyes, in particular luminescent metal complexes, and self-organizing materials, for application in displays (emissive, liquid crystalline, reflective), storage (optical, solid-state), electronics (mainly polymer-based) and sensors, e.g. for medical applications (diagnostics, imaging). Research on (opto-)electronic devices: preparation and characterization. Advanced instrumentation for ultra fast time-resolved measurements, for microscopy and for imaging, also for near-infrared luminescence.

-K. Brunner, J.A.E.H. van Haare, B.M.W. Langeveld-Voss, H.F.M. Schoo, J.W. Hofstraat, A. van Dijken, J. Phys. Chem. B, 106, 6834-6841 (2002).

-L.H. Slooff, A. van Blaaderen, A. Polman, G.A. Hebbink, S.I. Klink, F.C.J.M. van Veggel, D.N. Reinhoudt, J.W. Hofstraat, J. Appl. Phys., 91, 3955-3980 (2002).

Date submitted: 4th April 2006

Gerhard Holst, Ph.D.

Science & Research Department,
PCO AG,
Donaupark 11,
93309 Kelheim, Germany.
Tel: +49 (0)944 120 0536   Fax: +49 (0)944 120 0520
gerhard.holst@pco.de
www.pco.de

Specialty Keywords: CCD and CMOS Cameras, Luminescence Lifetime Imaging.

Main Interests: Research and development of high performance and high speed CCD and CMOS camera systems for the scientific and industrial applications, such as luminescence imaging, luminescence lifetime imaging, microscopy, time resolved spectroscopy, low light imaging, high speed imaging and optical chemical sensing.

-G. Holst, O. Kohls, I. Klimant, B. König, M. Kühl und T. Richter, 1998, A Modular Luminescence Lifetime Imaging System for Mapping Oxygen Distribution in Biological Samples, *Sensors and Actuators B,* Vol. 51, p. 163-170.

-G. Holst, I. Klimant, M. Kühl und O. Kohls, 2000, Optical Microsensors and Microprobes, in: Chemical Sensors in Oceanography, Ed. M. Varney, OPA Overseas Publisher Association, Amsterdam, p.143-188.

Date submitted: 29th September 2006

Graham Hungerford, Ph.D.

Departamento de Física,
Universidade do Minho,
4710-057 Braga,
Portugal.

graham@fisica.uminho.pt

Specialty Keywords: Sol-gel, Biosensor.
AIM 2005 = 9.9

My present research interests involve the manufacture and study (using fluorescence techniques) of sol-gel-derived matrices to elucidate dye-dye and dye-host interactions. This is oriented towards biosensor applications and the study of the interaction of proteins with biocompatble materials using fluorescent techniques.

-G. Hungerford et al. (2006). Use of fluorescence to monitor the incorporation of horseradish peroxidase into a sol-gel derived medium. *Biophys. Chem.* 120, 81-86.
-G. Hungerford et al. (2005). Studies on the interaction of nile red with horseradish peroxidase in solution. *FEBS. Journal* 272, 6161-6169.

Date submitted: 1st Sept. 2006

H. M. J. (Ria) Hut, Ph.D.

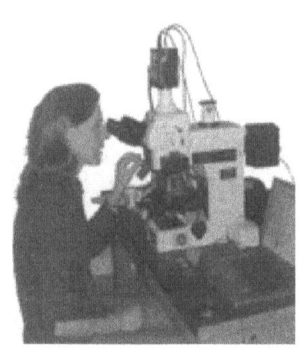

Lambert Instruments,
Turfweg 4,
9313 TH Leutingewolde,
The Netherlands.
Tel: +31 50 501 8461   Fax: +31 50 501 0034
rhut@lambert-instruments.com
www.lambert-instruments.com

Specialty Keywords: FLIM, fluorescence widefield microscopy.
My work is focused on contact with potential customers & to imply the needs of the biologists in the improvement of our system for Fluorescence Lifetime Imaging Microscopy (FLIM). There for I present our system during microscopy courses and conferences, and I'm experienced in Cell Biology by doing mainly fluorescence microscopy during my Ph.D. project at the Dept. of Radiation and Stress Cell Biology of the UMCG in The Netherlands. FLIM is mostly used to detect FRET between certain fluorescent proteins in cells. Examples of protein pairs suitable for FRET are GFP-RFP and CFP-YFP (donor-acceptor). The fluorescent proteins have to be linked to certain proteins of interest. Upon fluorescence lifetime changes of the donor, one can make conclusions whether the proteins are interacting with each other or not. Additionally, FLIM is used for pH measurements and ion imaging inside cells. More FLIM applications (or technology) can be found at our website.

Date submitted: 27th June 2005

Bernd Hutter, M.Sc.

Berthold Technologies GmbH. & Co. KG., Marketing,
Calmbacher Str. 22,
Bad Wildbad,
75323 Germany.
Tel: +49 7081 1770, fax +49 7081 177500
bernd.hutter@berthold.com
www.berthold.com/bio

Specialty Keywords: Fluorescence Microplate Readers,
Multimode Microplate Readers, Fluorescence Assays.

The current responsibilities at BERTHOLD TECHNOLOGIES include application and instrument handling support besides marketing issues. The knowledge is based on many years of experience with fluorescence, time-resolved fluorescence, fluorescence polarization, BRET and luminescence assays in academic research and pharmaceutical drug discovery.

-F. Berthold, C.H. Johnson, A. Heding, M. Peukert, M. Hennecke, B. Hutter (2002). Development of a sensitive instrument for Bioluminescence Resonance Energy Transfer (BRET) applications: Bioluminescence & Chemiluminescence: Progress & Current Applications, 193-196.

Date submitted: Editor Retained.

Takamitsu Ikkai, Ph.D.

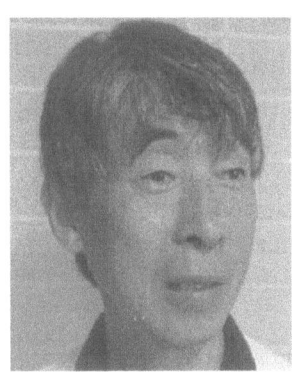

Biophysics, Aichi Prefectural University of Fine Arts,
Sagamine 1-1, Nagakute,
Aichi, 480-1194,
Japan.
Tel: 81 56 162 1180   Fax: 81 59 331 3406
ikkai@mail.aichi-fam-u.ac.jp
www.aichi-fam-u.ac.jp

Specialty Keywords: Excimer Fluorescence, Crystal, Actin.

If we want to know the dynamic mechanism of protein function, its structural change in solution based on the knowledge of crystal has to be studied. As a clue to this problem, we employed the excimer fluorescence which can be measured both in solution and crystal, and used pyrene-labeled actin as a sample. The structural dynamics monitored will bring a new information concerned with intramolecular rearrangement, not observed with other methods.

-T. Ikkai, K. Shimada (2002) Introduction of fluorometry to the screening of protein crystallization buffers. *J. Fluoresc* 12, 167-171.

Ito, A. S.
James, T. D.

Date submitted: Editor Retained.   **Amando S. Ito, Ph.D.**

Departamento de Física e Matemática,
FFCLRP – Universidade de São Paulo,
Av. Bandeirantes 3900, Ribeirao Preto,
14015 120, Brazil.
Tel: 55 16 302 3693
amando@dfm.ffclrp.usp.br
dfm.ffclrp.usp.br/fotobiofisica/membros.html

Specialty Keywords: FRET, Peptide Conformational Dynamics, Peptide / Lipid Interaction.

Research interests: Physico-chemical properties of extrinsic and intrinsic fluorescent probes for peptides and proteins. Donor-acceptor distance distribution and conformational dynamics in peptides. Labeled macromolecules in interaction with supramolecular assemblies. Fluorescence studies on membrane models.

-A.S. Ito, E.S. Souza, S.R. Barbosa and C.R. Nakaie. (2001) Fluorescence Study of Melanotropins Labeled with Aminobenzoic Acid. *Biophysical Journal*, 81, 1180-1189.
-D.C.Pimenta, I.L.Nantes, E.S.Souza, B. le Boniec, A.S.Ito, I.L.S.Tersariol, V.Oliveira, M.A.Juliano and L.Juliano. (2002) Interaction of heparin with internally quenched fluorogeic peptides. *Biochem. J.*, 366, 435-446.

Date submitted: 27th September 2006   **Tony D. James, Ph.D.**

Department of Chemistry,
University of Bath,
Bath, BA2 7AY,
UK.
Tel: +44 (0) 122 538 3810   Fax: +44 (0) 122 538 6231
t.d.james@bath.ac.uk
www.chemosensors.com

Specialty Keywords: Recognition, Boronic acid, Saccharide.
AIM 2005 = 20.7

Tony James is Senior Lecturer in Organic Chemistry at the University of Bath, heading the Boronic Acid Research Group. Research activities in his group cover the areas of supramolecular chemistry, sensor design, chiral recognition, saccharide recognition, anion recognition, synthetic organic chemistry, combinatorial chemistry and asymmetric synthesis.

-M. D. Phillips and T. D. James, (2004). Boronic acid based modular fluorescent sensors for glucose, J. Fluoresc. 14, 549-559.
- J. Zhao, M. G. Davidson, M. F. Mahon, G. Kociok-Köhn, and T. D. James (2004). An enantioselective fluorescent sensor for sugar acids, J. Am. Chem. Soc. 126, 16179-16186.

Date submitted: 4th July 2004

Andrzej Jankowski, Ph.D.

Unigversity of Zielona Gora,
Institute of Biotechnology and Environment Protection,
Monte-Cassino 21b. Zielona Gora,
Poland 65-561, Poland.
Tel: 048 071 353 9177
JJJ@WCHUWR.CHEJM.UNI.WROC.PL

Specialty Keywords: Fluorescence Spectroscopy, Photobiology, Environment Chemistry.

The main topics of scientific activity: 1) Structure of peptides and proteins. 2) Excited state proton transfer in proteins and in Langmuir - Blodgett films. 3) Photosensitization of bacteria.

-A. Mirończyk, A. Jankowski, A. Chyla, A. Ożyhar, P. Dobryszycki: Investigation of Excited State Proton Transfer Included in Langmuir-Blodgett Films. J. Phys. Chem. A 2004, 108, 5308-5314.

-A. Jankowski, S. Jankowski, A. Mirończyk: Synergistic Action of Photosensitizers and Normal Human Serum in a Bactericidal Process. Acta Microbiologica Polonica 2003, 52, 373-78.

Date submitted: 28th June 2005

Lennart B.-Å. Johansson, Ph.D.

Department of Chemistry,
Biophysical Chemistry,
Umeå University,
S-901 87 Umeå, Sweden.
Tel: +46 (0) 90 786 5149   Fax: +46 (0) 90 786 7779
Lennart.Johansson@chem.umu.se
staff.chem.umu.se/lennart.johansson/

Specialty Keywords: Electronic Energy Migration / Transfer, Protein Structure, Two-photon Excited Fluorescence.

Polarised one- and two-photon excited fluorescence is utilised to explore the structure of complex biomacromolecules. The project concerns theoretical modelling, experimental design and development of new versatile probes. The applications deal with fibrinolytic activity, pore-forming toxins, actin polymerisation, and amyloid diseases (*e.g.* BSE, Creutzfeld-Jacob´s and Alzheimer´s diseases).

-M. Isaksson, S. Kalinin, S. Lobov, S. Wang, M., T. Ny and L. B.-Å. Johansson. *Phys. Chem. Chem. Phys.* 6, 3001 (2004); P. Håkansson, M. Isaksson, P.-O. Westlund and L. B.-Å. Johansson, *J. Phys. Chem. B*, 108, 17243(2004).

Date submitted: 8th June 2005

Arthur E. Johnson, Ph.D.

Dept. Medical Biochemistry & Genetics,
Texas A&M University System Health Science Center,
116 Reynolds Medical Building, 1114 TAMU,
College Station, TX 77843-1114, USA.
Tel: 979 862 3188   Fax: 979 862 3339
ajohnson@medicine.tamhsc.edu
medicine.tamhsc.edu/wwrl
Specialty Keywords: Protein-Membrane Interactions, Protein
Trafficking, FRET.
AIM 2004 = 85.0

We are primarily investigating the movement of proteins through or into the ER or mitochondrial membranes (protein trafficking), and the creation of holes in mammalian cell membranes by bacterial cytolytic protein toxins. Molecular interactions and conformational changes involved in the assembly, function, and regulation of membrane-bound protein complexes are characterized by various fluorescence techniques. FRET is used to determine their structure and topography, to detect and quantify conformational changes, and to monitor intermolecular association.

-Woolhead, C.A., McCormick, P.J., and Johnson, A.E. (2004) Nascent membrane and secretory proteins differ in FRET-detected folding far inside the ribosome and in their exposure to ribosomal proteins. Cell 116, 725-736.

Date submitted: 14th June 2005

Carey K. Johnson, Ph.D.

Department of Chemistry,
University of Kansas,
1251 Wescoe Hall Drive, Lawrence,
KS 66045-7582, USA.
Tel: 785 864 4219   Fax: 785 864 5396
ckjohnson@ku.edu
www.chem.ku.edu/CJohnsonGroup/
Specialty Keywords: Single-molecule Fluorescence, Time-
Resolved Fluorescence.
AIM 2005 = 22.9

We investigate dynamics of proteins and peptides by single-molecule and time-resolved spectroscopy. We use single-molecule fluorescence spectroscopy and fluorescence correlation spectroscopy to investigate calcium signaling by single calmodulin molecules. Time-resolved fluorescence anisotropy and resonance energy transfer experiments are used to probe the dynamics and interactions of peptides and proteins.

-C.K. Johnson, K.D. Osborn, M.W. Allen, and B.D. Slaughter, (2005) Single-molecule fluorescence spectroscopy: new probes of protein function and dynamics, *Physiology* 20 (2), 10-14.

Date submitted: 28th June 2004

Michael L. Johnson, Ph.D.

Departments of Pharmacology and Internal Medicine,
University of Virginia Health System, Box 800735,
Charlottesville, VA 22908-0735,
U.S.A.
Tel: 434 924 8607   Fax: 434 982 3878
mlj8e@virginia.edu
mljohnson.pharm.virginia.edu

Specialty Keywords: Mathematical Modeling, Biophysics.

My research interests center on understanding the biochemical, physical chemical, thermodynamic, and physiological pathways by which one portion of a biological organism, or molecule, transfers information to other portions of the same organism, or molecule. My primary research tool is mathematical modeling.
Deconvolution Analysis as a Hormone Pulse-Detection Algorithm (2004) Johnson, M.L., Virostko, A., Veldhuis, J.D., and Evans, W.S., Methods in Enzymology 384, 40-53.

-Modulating the Homeostatic Process to Predict Performance During Chronic Sleep Restriction., (2004),. Johnson, M.L., Belenky, G., Redmond, D.P., Thorne, D.R., Williams, J.D., Hursch, S.R., Balkin, T.J., Aviation Space and Environmental Medicine 75, A141-A146.

Date submitted: Editor Retained.

Anita C. Jones, Ph.D.

School of Chemistry & Collaborative Optical Spectroscopy,
Micromanipulation and Imaging Centre,
University of Edinburgh,
Edinburgh, EH9 3JJ, UK.
Tel: +44 (0) 131 650 6449   Fax: +44 (0) 131 650 4743
a.c.jones@ed.ac.uk
www.chem.ed.ac.uk and www.cosmic.ed.ac.uk

Specialty Keywords: Spectroscopy, Photophysics, Time-resolved Fluorescence, FLIM.

Research interests: Steady state and time-resolved fluorescence spectroscopy; fluorescence lifetime imaging; molecular photophysics and photochemistry; use of fluorescence to probe biomolecular systems; photophysics of luminescent polymers; industrial and biomedical applications of fluorescence.

-N.M. Speirs , W.J. Ebenezer and A.C. Jones (2002). Observation of a fluorescent dimer of a sulfonated phthalocyanine, *Photochem.Photobiol* 76, 247-251.
-A C Jones, M. Millington, J Muhl, J M De Freitas, J S Barton and G Gregory (2001). Calibration of an optical fluorescence method for film thickness measurement, *Measurement Science & Technology*, 12, N23-N27.

Date submitted: 17th October 2006

Pedro A. S. Jorge, Ph.D.

Optoelectronics Unit, INESC Porto,
Rua do Campo Alegre, 687,
4169-007, Porto,
Portugal.
Tel: +35 122 608 2601   Fax: +35 122 608 2799
pjorge@inescporto.pt
www.inescporto.pt

Specialty Keywords: Optical fiber, Biochemical, Sensors.
AIM 2006 = 2.7

From 1997 Pedro Jorge is a researcher at INESC Porto working in the field of optical fiber sensors. In the framework of his PhD he developed work in luminescence based oxygen and pH sensors for medical and environmental applications. More recently, in collaboration with the University of North Carolina at Charlotte, USA, he has been exploring the potential applications of semiconductor nanoparticles in the field of optical fiber biochemical sensing.

-P.A.S.Jorge, M.Mayeh, R.Benrashid, P.Caldas, J.L.Santos, and F.Farahi (2006). Applications of quantum dots in optical fiber luminescent oxygen sensors *Appl. Optics* 45(16), 3760-3767.
-P.A.S.Jorge, M.Mayeh, R.Benrashid, et al. (2006). Quantum dots as self-referenced optical fibre temperature probes for luminescent chemical sensors *Meas. Sci. Technol.* **17**(5), 1032-1038.

Date submitted: 29th September 2006

Nicoletta Kahya, Ph.D.

Philips Research Laboratories,
Molecular Diagnostics,
High Tech Campus 4, Eindhoven,
5656 AE, The Netherlands.
Tel: +31 40 274 3794   Fax: +31 40 274 3350
Nicoletta.kahya@philips.com

Specialty Keywords: Membrane, Microdomains, fluorescence correlation spectroscopy.
AIM 2005 = 39.0

My research interest focuses on lipid and protein spatio-temporal organization in both cellular and artificial membranes, in order to better understand the links between the structural membrane architecture and the function of the embedded membrane components. Lipid-protein and protein-protein interactions are tracked mainly by fluorescence microscopy techniques.

-N. Kahya, D. Scherfeld, K. Bacia and P. Schwille (2003). Probing lipid mobility of raft-exhibiting model membranes by fluorescence correlation spectroscopy *J. Biol. Chem.* 278(30), 28109-28115.
-L. Kalvodova, N. Kahya, P. Schwille, R. Ehehalt, P. Verkade, D. Drechsel and K. Simons (2005). Lipids as modulators of proteolytic activity of BACE: involvement of cholesterol, glycosphingolipids, and anionic phospholipids in vitro *J. Biol. Chem.* 280(44), 36815-36823.

Date submitted: 18th August 2005

Inta Kalniņa, M.D.

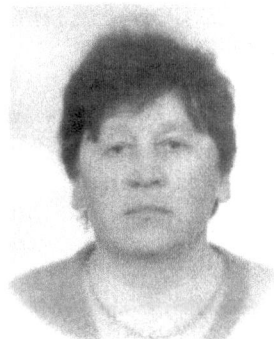

Dep. Of Organic Chemistry,
1 Kalku Str., Riga,
LV-1583,
Latvia.
Fax: 371 722 5039
lulc@lanet.lv

Specialty Keywords: Fluorescent Probes, Lymphocytes, Diagnostics.

AIM 2004 = 0.8

Reseach interests are: 1. Testing of new fluorescent dyes with the aim of its application in medical diagnostics, 2. Development of new methods and its application in clinics, using a newly sinthesized probes: derivatives of 3-aminobenzanthrone (ABM), naphthalic acid and styrilpyridinium. Observed patients was with several nonmalignant (advanced lung tuberculosis, multiple sclerosis, rheumatoid arthritis, ischemic heart diseases etc.), malignant diseases (gastrointestinal cancer, B-cell lynphoid leukemia etc.) and those who have been subjected to ionizing radiation during the clean–up work in Chernobyl. 3. Using fluorescent probe methods for investigations of cellular mechanisms of immunity regulation.

-R. Bruvere, N. Gabruševa, I. Kalnina, G. Feldmane, I. Meirovics. (2003).Fluorescent characteristics of blood leukocytes of patients with malignant and nonmaliganant diseases. J. Fluoresc. 13 (2), 149-156.

Date submitted: 9th May 2006

Ray Kaminski, Ph.D.

VP of Fluorescence,
HORIBA Jobin Yvon,
3880 Park Avenue,
Edison, NJ 08820-3012, USA.
Tel: 732 494 8660 Ext: 121   Fax: 732 549 5157
Ray.Kaminski@JobinYvon.com
www.jobinyvon.com

Specialty Keywords: Fluorescence Instruments, Phosphorescence Instruments, Lifetime Instruments.

Dr. Kaminski's current interests include: Beta-decay of pions at Los Alamos Meson Physics Facility.   Work with phosphorescence of uranyl, remote sensing with portable fluorometers, as well as multi-frequency phase fluorimetery.

Date submitted: 6th April 2006

Hee Chol Kang, Ph.D.

Technical Area Manager,
Dye Chemistry R/D, Invitrogen Corporation,
29851 Willow Creek Road, Eugene,
Oregon, 97402, USA.
Tel: 541 335 0342   Fax: 541 335 0206
heechol.kang@invitrogen.com
www.invitrogen.com

Specialty Keywords: Fluorescence, Time-resolved Probes, Nucleotides.

My research focuses on the design and development of novel fluorescent molecules with improved spectral properties. Recent projects include: Design and synthesis of novel fluorescent organometallic complexes with long fluorescence lifetimes and high Stokes shift for time-resolved applications, design and synthesis of fluorescent probes for direct chemical labeling of nucleic acids, and, synthesis of a wide variety of fluorescent nucleotides for the study of nucleotide-binding proteins.

Date submitted: 23rd October 2006

Peter Kapusta, Ph.D.

PicoQuant GmbH.,
Rudower Chaussee 29,
12489 Berlin,
Germany.
Tel: +49 306 392 6914
kapusta@pq.fta-berlin.de
www.picoquant.com

Specialty Keywords: Pulsed Diode Lasers, LED, Time-Resolved Spectroscopy, Single Molecule Detection, FCS.

Main interests: development of laser diode and LED based time-resolved fluorescence instrumentation, promotion of the TCSPC method in various research areas, ultrasensitive detection including SMD, photophysics of fluorophores, energy and charge transfer in molecules, solvation dynamics.

-Kapusta P., Wahl M., Benda A., Hof M., Enderlein J. (2006), *J. of Fluorescence*, in press.
Benda A., Hof. M., Wahl M., Patting M., Erdmann R., Kapusta P. (2005), *Rev. Sci. Instrum.*76, 033106.
-Kraemer B., Koberling F., Ortmann U., Wahl M., Kapusta P., Buelter A., Erdmann R.(2005), Proc. of SPIE, 5700, 138.

Date submitted: 10th August 2005

Murad Karmali.

The COOKE Corporation,
6930 Metroplex Drive,
Romulus, MI 48174,
USA.
Tel: 248 276- 820 ext 212   Fax: (248) 276 8825
murad.karmali@cookecorp.com
www.cookecorp.com

Specialty Keywords: CCD and CMOS Cameras, em CCD Cameras.

Main Interests: Development and production of high performance and high speed CCD and CMOS camera systems for the scientific and industrial applications, such as LIF, microscopy, Chemiluminescence, time resolved spectroscopy, combustion imaging and PIV imaging.

Date submitted: 25th September 2004

Jan Karolin, Ph.D.

Department of Physics,
University of Strathclyde,
107 Rottenrow, Glasgow, G4 ONG,
Scotland, UK.
Tel: +44 (0)141 548 4012
jan.karolin@strath.ac.uk

Specialty Keywords: Time-resolved Fluorescence, Multiphoton Induced Fluorescence, Resonance Energy Transfer.

Current research is focused on the development of fluorescence spectroscopic techniques to characterize morphological properties of silica materials synthesized through sol-gel routes, i.e. through room temperature wet chemical approaches. With structural features on a nanometer length scale we are looking to correlate observations from fluorescence depolarization, energy transfer and solvent relaxation to parameters such as pore volume and surface area.

-C.D. Geddes, J. Karolin and D. J. S. Birch (2002). 1 and 2-photon fluorescence anisotropy decay in silicon alkoxide sol-gels: *J. Phys. Chem. B*, 106(15), 3835-3841.
-J. Karolin, C.D. Geddes, K. Wynne and D. J. S. Birch (2002) Nanoparticle metrology in sol-gels using multiphoton excited fluorescence: *Meas. Sci. Technol.* 13(1), 21-27.

Karuso, P.
Kaspar, L.

Date submitted: Editor Retained.

Peter Karuso, Ph.D.

Department of Chemistry,
Macquarie University,
Sydney, NSW, 2109,
Australia.
Tel: +612 9 850 8275   Fax: +612 9 850 8313
peter.Karuso@mq.edu.au
www.chem.mq.edu.au/~vislab

Specialty Keywords: Proteomics, Natural Products, Bioassay.

The Karuso group specialize in natural products chemistry and the discovery/application of new fluorescent technologies. Past achievements include a fluorescence based antimicrobial assay[1] and the isolation of a fluorescent natural product from a fungus[2] that is being developed as a powerful 2D gel electrophoresis stain. Current interests include the isolation of new fluorescent stains and the synthesis of fluorescence based molecular rulers.

-S. Chand, I. Lusunzi, L. R. Williams, D. A. Veal and P. Karuso (1994) Rapid screening of the antimicrobial activity of extracts and natural products. *J. Antibiotics* 47, 1295–1304.
-P. J, L. Bell and P. Karuso (2003) Epicocconone, a novel fluorescent compound from the fungus *Epicoccum nigrum. J. Amer. Chem. Soc.* 125, 9304–9305.

Date submitted: 10th August 2005

Luitpold Kaspar.

PCO AG,
Sales Department,
Donaupark 11, 93309 Kelheim,
Germany.
Tel: +49 (0)944 120 0550   Fax: +49 (0)944 120 0520
luitpold.kaspar@pco.de
www.pco.de

Specialty Keywords: CCD and CMOS Cameras, em CCD Cameras.

Main Interests: Development and production of high performance and high speed CCD and CMOS camera systems for the scientific and industrial applications, such as particle image velocimetry (PIV), microscopy, fluorescence spectroscopy, combustion and spray analysis and ultra speed imaging.

Date submitted: 3rd October 2006

Demet Kaya Aktaş, Ph.D.

Department of Physics,
Istanbul Technical University,
Maslak, Istanbul,
34469, Turkey.
Tel: +90 212 285 6603   Fax: +90 212 285 6386
demet@itu.edu.tr
www.fizik.itu.edu.tr/kayad/
Specialty Keywords: Polymeric gels, Critical phenomena,
Universality.
AIM 2004 = 4.8

I have been studying critical phenomena in different polymeric gels by using steady-state and / or time-resolved fluorescence techniques. These techniques made it possible to study the glass transition in the various bulk gels and showed that this transition is in the same universality class as percolation. On the other hand, in hydrogels a crossover was obtained from percolation to mean field values depending on the monomer concentration.

-D.Kaya, Ö.Pekcan and Y. Yılmaz *Phy. Rev. E,* 69,16117 (2004)

-Ö.Pekcan and D.Kaya *Composite Interfaces* 12, 6, 501 (2005)

Date submitted: 16th June 2005

Klaus K. Kemnitz, Ph.D.

EuroPhoton GmbH.,
Mozartstr. 27,
Berlin D-12247,
Germany.
Tel: +49 30 7719 0145   Fax: +49 30 771 4450
klauskemnitz@aol.com
www.europhoton.de

Specialty Keywords: FLIM & FLIN, TSCSPC, Minimal-Invasive.

K.K. has 25 years experience in picosecond micro-spectroscopy and founded EuroPhoton in 1995. RTD of ultra-sensitive imaging systems for spectroscopy and microscopy (FLIM & FLIN) and development of the TSCSPC method for application in cell biology, using non-scanning DL-/QA- imaging detectors for minimal-invasive, multi-parametric observation of living cells and single molecules. Patents in the field of multi-parameter FLIM & FLIN, minimal-invasive FRET Verification, and Long-Period Observation (LPO) of Single Molecules and Motors. EC-project coordinator of SingleMotor-FLIN (www.euro-flin.de).

-K. Kemnitz, New Trends in Fluorescence Spectroscopy: Methods and Applications, Eds. B. Valeur and J.-C. Brochon, Springer, 1(2001)381.

-M. Tramier, K. Kemnitz, C. Durieux, M. Coppey-Moisan, J. Microscopy, 213(2004)110.

Date submitted: 4th April 2006

Norbert U. Kemnitzer, Ph.D.

ATTO-TEC GmbH.,
Am Eichenhang 50,
D-57076 Siegen,
Germany.
Tel: +49 (0) 271 740 4735
norbert.kemnitzer@gmx.de & kemnitzer@atto-tec.com
www.atto-tec.com

Specialty Keywords: Organic Dyes, Fluorescent Labels.

My research interest is the development of fluorescent dyes suitable as labels for applications in biochemistry and medicine. I am particularly interested in the design of chromophoric systems for the red region of the visible spectrum. Another important field of my work is to reduce dye-aggregation in aqueous media by attaching hydrophilic groups to the chromphor.

-J. Arden-Jacob, J. Frantzeskos, N.U. Kemnitzer, A. Zilles, and K.H. Drexhage (2001). New fluorescent markers for the red region *Spectrochim. Acta A* 57(11), 2271-2283.
-N.U. Kemnitzer (2001). Ph.D. Thesis. Amidopyrylium-Fluoreszenz-Farbstoffe. Der Andere Verlag, Osnabrück. Available online at: http://www.ub.uni-siegen.de/epub/diss/kemnitzer.htm

Date submitted: 1st April 2005

Edward Kiegle, Ph.D.

Chroma Technology Corp.,
10 Imtec Lane,
Rockingham, Vermont 05101,
USA.

Tel: 800 824 7662 toll free / 802 428 2500   Fax: 802 428 2525
ekiegle@chroma.com
www.chroma.com

Specialty Keywords: Fluorescent Proteins, Transcriptional Profiling, Plant Genomics.

My background is in plant molecular biology and functional genomics, relying heavily on fluorescent and bioluminescent reporters and transcriptome profiling in rice and Arabidopsis. Currently I provide technical support for fluorescent protein, FRET, and functional genomic applications for Chroma Technology Corp.

Date submitted: 31ˢᵗ August 2005

Borys Kierdaszuk, Ph.D., D.Sc.

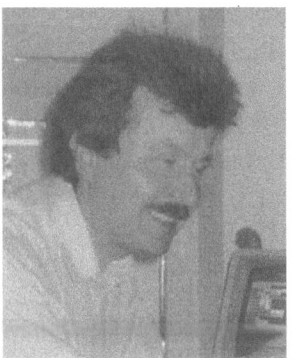

University of Warsaw, Department of Biophysics,
Institute of Experimental Physics,
PL-02089 Warsaw,
Poland.
Tel: +48 22 554 0784   Fax: +48 22 554 0771
borys@biogeo.uw.edu.pl
www.biogeo.uw.edu.pl/people/borys_en.html

Specialty Keywords: Interpretation of Emission Decays.

Mechanism of action of biological molecules (e.g. enzymes) with the aid of emission spectroscopy (fluorescence, phosphorescence), including interpretation of emission decays.

-J. Włodarczyk, and B. Kierdaszuk, (2003), *Biophys. J.* 85, 589-598.
-J. Włodarczyk, and B. Kierdaszuk, (2004), *Chem. Phys.* 297, 139-142.
-J. Włodarczyk, G. S. Galitonov, and B. Kierdaszuk, (2004), *Eur. Biophys. J.* 33, 377-385.
-J. Włodarczyk, and B. Kierdaszuk, (2004), Progress in Biomedical Optics and Imaging, Vol. 5. Complex Dynamics, Fluctuations, Chaos and Fractals in Biomedical Photonics, Proceedings of SPIE, Vol. 5330, pp. 92-100.
-J. Włodarczyk, and B. Kierdaszuk, (2005), *Acta Phys. Polon.* 107, 883-894.

Date submitted: 29ᵗʰ March 2006

Paavo K. J. Kinnunen, Ph.D.

Memphys – Center for Biomembrane Physics,
Helsinki Biophysics & Biomembrane Group,
Department of Medical Chemistry, Institute of Biomedicine,
POB 63, FIN-00014, University of Helsinki, Finland.
Tel: +35 891 912 5400   Fax: +35 891 912 5444
Paavo.Kinnunen@Helsinki.Fi

Specialty Keywords: Lipids, Biomembranes, Lipid-protein Interactions.

The major line of research of HBBG pursues the molecular level mechanisms underlying both 2-D and 3-D ordering of supramolecular assemblies constituted by lipids, aiming to compile an integrated view on the coupling between the physical properties of lipids to the physiological functions of biomembranes. More specifically, we are elucidating the mechanisms which convey changes in the physicochemical characteristics of bilayer lipids to the conformation and activity of membrane proteins.

Kirsch-De Mesmaeker, A.
Kleszczyńska, H.

Date submitted: Editor Retained.

Andrée Kirsch-De Mesmaeker, Ph.D.

Organic Chemistry and Photochemistry,
Université Libre de Bruxelles, Faculty of Sciences,
CP 160/08, 50 av. F.D. Roosevelt,
1050 Brussels, Belgium.
Tel: 32 (0)2 650 3017   Fax: 32 (0)2 650 3606
akirsch@ulb.ac.be
www.ulb.ac.be/sciences/cop

Specialty Keywords: Ru(II) Complexes, DNA, Dendrimer.

Interaction and photoreaction of Ru(II) and Rh(III) complexes with DNA, examined by spectroscopic methods and gel electrophoresis analyses.   Study of Ru(II) derivatized oligonucleotides in the frame of the antisense and antigene strategy or as molecular tools in genes analysis.   Ru-induced photocrosslinking of oligonucleotides.   Preparation and study of polynuclear Ru(II) complexes and dendrimers for applications with biomolecules or as antenna systems for the collection of light.

-O. Lentzen, .F. Constant, E. Defrancq, M. Prevost, S. Schumm, C. Moucheron,
P. Dumy, A. Kirsch-DeMesmaeker, ChemBioChem, 4 (2003), 195-202.
-C. Moucheron, A. Kirsch-DeMesmaeker, A. Dupont, E. Leize, A. Van Dorsselaer,
J. Am. Chem. Soc., 118 (1996), 12834-12835.

Date submitted: 28th June 2006

Halina Kleszczyńska, Ph.D.

Department of Physics and Biophysics,
Agricultural University,
50-375 Wrocław, Norwida 25,
Poland.
Tel: +48 71 320 5141   Fax: +48 71 320 5172
halina@ozi.ar.wroc.pl
www.ar.wroc.pl

Specialty Keywords: Erythrocyte, Membrane fluidity,
Biological active compounds.
AIM 2006 = 6.6

Our research activity concerns new synthesized membrane-active compounds for application as pesticides or free radical scavengers. Their potential activity is tested with the use of model and biological membranes and plants. One of the techniques used is fluorescence. Investigations permit to determine the potential biological activity of these compounds, the mechanism of their interaction with mentioned objects and the role structural features of the compounds play in this interaction. Data obtained may be helpful while synthesizing new biological active compounds.

-Kleszczyńska H., Bonarska D., Pruchnik H., Bielecki K., Piasecki A., Łuczyński J. Sarapuk J. (2005) Z. Naturforsch. 60c, 667-671.
-Kleszczyńska H., Bonarska D., Łuczyński J, Witek S., Sarapuk J. (2005) J. Fluorescence 15 (2) 137-141.

Date submitted: 23ʳᵈ October 2006

Alex E. Knight, Ph.D.

National Physical Laboratory

Biotechnology Group, Quality of Life Division,
National Physical Laboratory,
Hampton Road, Teddington,
Middlesex, TW11 0LW, UK.
Tel: +44 (0) 20 8943 6308
alex.knight@npl.co.uk
www.npl.co.uk/biotech

Specialty Keywords: Single molecule fluorescence,
Fluorescence standards, Biomolecule spectroscop.

I am interested in the applications of single molecule detection – particularly single molecule fluorescence – to the life sciences. I also have interests in fluorescence standards for the life sciences and biological application of spectroscopy.

-A. Knight, G. Mashanov, and J. Molloy (2005). Single molecule measurements and biological motors *European Biophysics Journal* 35(1), 89.
-J. A. Gaunt, A. E. Knight, S. A. Windsor, and V. Chechik (2005). Stability and quantum yield effects of small molecule additives on solutions of semiconductor nanoparticles *J Colloid Interface Sci* 290(2), 437-443.

Date submitted: Editor Retained.

Karsten König, Ph.D.

Center for Lasermicroscopy, University Jena,
Teichgraben 7,
Jena, 07743,
Germany.
Tel: +49 364 193 8560   Fax: +49 364 193 8590
kkoe@mti-n.uni-jena.de
www.uni-jena.de/clm

Specialty Keywords: Multiphoton Microscopy, Time-resolved Single Photon Counting, Tissue Imaging, Autofluorescence.

Research is focussed on multiphoton fluorescence microscopy and imaging of tissue autofluorescence with high submicron spatial resolution, 250ps temporal resolution and 5nm spectral resolution. Our studies include the single/few molecule level (e.g. Multiphoton Multicolor FISH, time-resolved FRET), the single cell level (e.g. GFP expression after optical gene transfer, imaging of optically trapped gametes and microorganisms) and *in vivo* studies on tissues (optical multiphoton tomography of skin and eyes). The equipment includes femtosecond laser scanning microsopes, a TauMap microscope for fluorescence lifetime imaging, systems for nanosurgery and imaging, laser tweezers and the multiphoton skin imaging system DermaInspect 100.

-König: Review. Multiphoton microscopy in life sciences. J. Microsc. 200(2000)83-104.
-Tirlapur, König: Targeted transfection by femtosecond laser. Nature. 418(2002)290-291.

Date submitted: 29th June 2005

Vladyslava B. Kovalska, Ph.D.

Inst. of Molecular Biology and Genetics of NAS of Ukraine,
Zabolotnogo Str. 150,
Kyiv, 03143,
Ukraine.
Tel: / Fax: +38 044 522 2458
yarmoluk@imbg.org.ua
www.yarmoluk.org.ua

Specialty Keywords: Fluorescent Probes, Cyanine Dyes,
Nucleic Acids.

The research activities of Dr. V. Kovalska are aimed on the designing of fluorescent probes for nucleic acid and protein detection [1]. Now she is working at the Department of Combinatorial Chemistry of Biological Active Compounds under the guiding of Dr. S. Yarmoluk. Her present researches are devoted to the characterization and studies of mechanism of fluorescent cyanine dyes – biopolymers interaction with the use of spectral-luminescent methods [1, 2].

-V.B. Kovalska, M.Yu. Losytskyy, D.V. Kryvorotenko, A.O. Balanda, V.P. Tokar, S.M. Yarmoluk (2005) Dyes and Pigments, 68, 39-45.
-V.B. Kovalska, M.Yu. Losytskyy and S.M. Yarmoluk (2004) Spectrochimica Acta Part A: Molecular and Biomolecular Spectroscopy, 60, 129-136.

Date submitted: 19th July 2004

Arjen A. M. Krikken, Ing.

Eukaryotic Microbiology, University of Groningen,
Kerklaan 30, Haren,
Groningen, 9751NN,
The Netherlands.
Tel: +31 0 50 363 2177   Fax: +31 0 50 363 2154
krikkena@biol.rug.nl
www.rug.nl/gbb/research/researchGroups/
eukaryoticMicrobiology/

Specialty Keywords: Peroxisome, Yeast.

Central theme of the research in the group Eukaryotic Microbiology is to study the relationships that exist between the structure and function of eukaryotic cells/cellorganelles. During the past fifteen years the work has been focussed on the principles of the homeostasis (biogenesis and selective turnover) and metabolic functioning of microbodies (peroxisomes, glyoxysomes) in yeast and in filamentous fungi.

Date submitted: 27th June 2004

Mikael Kubista, Ph.D.

TATAA Biocenter,
Medicinarg. 7B, Goteborg,
405 30,
Sweden.
Tel: +46 31 7733926   Fax: +46 31 7733948
Mikael.kubista@tataa.com
www.tataa.com

Specialty Keywords: Multidimensional Spectroscopy,
Fluorescent Probes, Realtime PCR.

Our research interest spans from characterization of molecular interactions by multidimensional spectroscopy to the development of dye and fluorescent probes for nucleic acid detection. Our most important contribution to the area of life sciences is the LightUp probe for sequence specific detection of nucleic acids in homogeneous solution. Presently, we are developing technology to measure gene expression in individual cells using a real-time PCR microchip.

Date submitted: 10th October 2006

Alexander V. Kukhta, Ph.D.

Sector of Electron Spectroscopy and Optics,
Institute of Molecular & Atomic Physics, NAS of Belarus,
Nezalezhnastsi Ave. 70, Minsk, 220072,
Belarus.
Tel: +37 517 284 1719   Fax: +37 517 284 0030
Kukhta@imaph.bas-net.by

Specialty Keywords: Electron-molecular Interaction, Organic
Cathodo- and Electroluminescence, Charge Transport.
AIM 2005 = 4.7

Luminescent properties of biological and electroactive organic molecules under irradiation by low-energy monokinetic electrons with variable energies from 0 to 100 eV. Physics of electron-molecular interactions and transport of electrons through different ordered and disordered organic media. Electroluminescence properties of low-molecular weigth and polymer materials, new electroluminescent materials and structures.

-A.V.Kukhta, E.E.Kolesnik, I.K.Grabchev, S.A.Sali (2006) Spectral and Luminescent Properties and Electroluminescence of Polyvinylcarbazole with 1,8-Naphthalimide in the Side Chain *J. Fluorescence* 16 .

-A.V. Kukhta (2005). Transport of low-energy electrons in thin organic films *LiqCryst.Mol.Cryst.* 427, 71-93.

Kürner, J. M.
Kusumi, A.

Date submitted: Editor Retained.

Jens M. Kürner, Ph.D.

Competence Center for Fluorescent Bioanalysis,
Josef-Engert-Straße 9,
D – 93053 Regensburg,
Germany.
Tel: +49 (0)941 943 5011   Fax: +49 (0)941 943 5018
jens.kuerner@exfor.uni-regensburg.de
www.kfb-regensburg.de

Specialty Keywords: Array-technology, Biotechnology,
Microscopy / Flow-Cytometry, Synthesis / Spectroscopy.

The Competence Center for Fluorescent Bioanalysis, which is affiliated to the University of Regensburg and is located in the BioPark Regensburg building, is a competent service provider of customer-oriented research and development. In addition to providing the diagnostic tools for research and development in pharmaceutical companies, the competence center focuses on customers in national and international biotechnology companies as well as private and public research institutes. The objective is to offer interdisciplinary research and development services in fluorescent bioanalysis in a unique network. This concept is based on the integration of components in chemistry, biology, medicine and engineering sciences through the utilization of the facilities for research and developement at the University of Regensburg, the University of Applied Sciences of Regensburg and the University Hospital of Regensburg.

Date submitted: Editor Retained.

Akihiro Kusumi, D.Sc.

Department of Biological Science,
Nagoya University Chikusa-ku,
Nagoya, 464 8602,
Japan.
Tel: +81 52 789 2969   Fax: +81 52 789 2968
akusumi@bio.nagoya-u.ac.jp
www.supra.bio.nagoya-u.ac.jp

Specialty Keywords: Single Molecules, Cell Membrane, Signal Transduction.

We develop single molecule techniques to be used for the study of live cells, such as single particle tracking and single fluorophore video imaging of membrane proteins, and single molecule dragging of membrane molecules using optical traps. Using these technologies, we study the mechanisms of signal transduction in the cell membrane, development of neuronal network, interaction of the membrane skeleton with membrane molecules, and formation and the functional mechanism of rafts, caveolae, and coated pits.
-T. Fujiwara, K. Ritchie, K. Metz-Honda, K. Jacobson, and A. Kusumi. Phospholipids undergo hop diffusion in compartmentalised cell membrane. J. Cell Biol. 157, 1071-1081 (2002).
-R. Iino, I. Koyama, and A. Kusumi. Single molecule imaging of GFP in living cells: E-cadherin forms oligomers on the free cell surface. Biophys. J. 80, 2667-2677 (2001).

Date submitted: 24th May 2006

Alexey S. Ladokhin, Ph.D.

University of Kansas Medical Center,
Department of Biochemistry and Molecular Biology,
3901 Rainbow Blvd., Kansas City,
KS 66160-7421, USA.
Tel: 913 588 7006   Fax: 913 588 7007
aladokhin@kumc.edu
www.kumc.edu/biochemistry/Ladokhin.html

Specialty Keywords: Membrane Protein Insertion / Folding,
Depth-dependent Quenching, Red-edge Effects.

My research focuses on understanding the structural and thermodynamic principles of insertion and assembly of membrane proteins and uses fluorescence spectroscopy as a principal tool. Over the years we have developed and applied fluorescence methods enabling us to characterize the depth of membrane penetration into the bilayer, the lipid exposure and cis/trans topology of particular sites as well as the conformational heterogeneity of membrane-inserted proteins and peptides.

-Palchevskyy *et al.* (2006). Chaperoning of membrane protein insertion into lipid bilayers by hemifluorinated surfactants: application to diphtheria toxin *Biochemistry* 45, 2629-2635.

-Posokhov and Ladokhin (2006)   Lifetime fluorescence method for determining membrane topology of proteins, *Analytical Biochem.* 348, 87-93.

Date submitted: 7th April 2006

Joseph R. Lakowicz, Ph.D.

Center for Fluorescence Spectroscopy,
Dept. of Biochemistry and Molecular Biology,
University of Maryland School of Medicine,
725 West Lombard St, Baltimore, Maryland, 21201, USA.
Tel: 410 706 7978/8409   Fax: 410 706 8408
lakowicz@cfs.umbi.umd.edu
http://cfs.umbi.umd.edu

Specialty Keywords: Fluorescence.
AIM 2006 = 41.6

Current Research Interests: My research is focused on advancing the field of fluorescence spectroscopy. At present my interests are focused on developing the use of metallic nanostructures on nearby fluorophores, which interact with plasmons on the metals. We refer to this topic as Plasmon-Controlled Fluorescence.

-Radiative decay engineering 5: Metal-enhanced fluorescence and plasmon emission. 2005,J. R. Lakowicz, Analytical Biochemistry 337,171-194.

-Surface-plasmon-coupled emission of quantum dots, J. R. Lakowicz et al. J.Phys. Chem. B 2005, 109,1088-1093.

Date submitted: 24th April 2006

Marek Langner, Ph.D.

Department of Physics,
Wrocław University of Technology,
Wyb. Wyspiańskiego 27, Wrocław,
50-370, Poland.
Tel: +48 71 320 2384   Fax: +48 71 328 3696
marek.langner@pwr.wroc.pl

Specialty Keywords: Membrane Biophysics, Targeted Drug
Delivery Systems, Supramolecular Aggregates.
AIM 2005 = 7.7

Application of fluorescence techniques to study the topology and formation of supramolecular aggregates, designed for the targeted delivery of biologically active compounds. This includes aggregates containing macromolecules (DNA, proteins) and drug-lipid ensembles. Studies on the molecular level are performed, as well as studies of cooperative processes occurring in the lipid bilayer when driven out of equilibrium. The cooperativity of the system is analyzed based on relaxation processes.

-K Bryl, M. Langner (2005)"Fluorescence applications in targeted drug delivery." In "Fluorescence spectroscopy in biology. Advanced methods and their applicaitons to membranes, proteins, DNA, and cells." Ed. M. Hof, R Hutterer, V Fidler Sringer-Verlag Berlin, Heidelberg, pp. 229-242.

Date submitted: 10th October 2006

Thomas M. Laue, Ph.D.

Biochemistry and Molecular Biology,
U. New Hampshire Rudman Hall, 46 College Road,
Durham, NH 03824, USA.
(Center to Advance Molecular Interaction Science (CAMIS))
(Biomolecular Interaction Technologies Center (BITC))
Tel: (603) 862 2459   Fax: (603) 862 0031
tom.laue@unh.edu
www.camis.unh.edu & www.bitc.unh.edu

Specialty Keywords: Fluorescence Optics, Analytical
Ultracentrifuge, Binding Strength and Characterization.

CAMIS develops unique instruments to characterize molecular interactions such as a fluorescence detector for the AUF. BITC is an NSF Industry/University Cooperative Research Center composed of global pharmaceutical firms and instrument manufacturers.

-Laue, T.M., Austin, J.B., Rau, D.A. (2006). "A Light Intensity Measurement System for the Analytical Ultracentrifuge." Progr Colloid Polym Sci., 131, 1-8.
-MacGregor, IK, Anderson AL, Laue TM (2004) "Fluorescence Detection for the XLI Ultracentrifuge" Biophys. Chem, 108, 165-185.

Date submitted: 13th October 2006

Dušan Lazár, Ph.D.

Laboratory of Biophysics, Faculty of Science,
Palacký University,
tř. Svobody 26, 771 46 Olomouc,
Czech Republic.
Fax: +42 058 522 5737
lazard@seznam.cz
exfyz.upol.cz/bf

Specialty Keywords: Chlorophyll fluorescence, Model.

My research interest is biophysics of photosynthesis. I mainly use chlorophyll fluorescence techniques, particularly the so-called fluorescence induction. I am further interested in mathematical modeling of the fluorescence induction and other photosynthetic quantities as well as in fluorescence imaging and statistical evaluation of fluorescence parameters.
-D. Lazár, P. Sušila, and J. Nauš (2006). Early Detection of Plant Stress from Changes in Distributions of Chlorophyll *a* Fluorescence Parameters Measured with Fluorescence Imaging. *J. Fluoresc.* 16, 173-176.; D. Lazár (2003).
-Chlorophyll *a* fluorescence rise induced by high light illumination of dark-adapted plant tissue studied by means of a model of photosystem II and considering photosystem II heterogeneity. *J. Theor. Biol.* 220, 469-503.

Date submitted: 23rd March 2005

Robert P. Learmonth, Ph.D.

Centre for Rural & Environmental Biotechnology,
Department of Biological and Physical Sciences,
University of Southern Queensland,
Toowoomba, QLD 4350, Australia.
Phone: +61 74 631 2361   Fax: +61 74 631 1530
learmont@usq.edu.au
www.usq.edu.au/biophysci

Specialty Keywords: Multi-photon Microscopy, Membrane Fluidity, Yeast.

Research areas: yeast biotechnology, cell membrane biochemistry/biophysics, fluorescence spectroscopy and microscopy. Using yeast as a model system to investigate how cells react to changes in environment, focusing on cell membranes as the critically important structures in adaptation. Development of methods using novel fluorescent probes and multi-photon microscopy to study membrane status in single cells of yeasts, bacteria and other microbes.

-Croney, J.C., Jameson, D.M. and Learmonth, R.P. (2001) Fluorescence Spectroscopy in Biochemistry: Teaching Basic Principles with Visual Demonstrations. *Biochemistry and Molecular Biology Education* 29, 60-65.

Date submitted: Editor Retained.

W. Jonathan Lederer, M.D., Ph.D.

Medical Biotechnology Center,
University of Maryland Biotechnology Institute,
725 W. Lombard Street, Baltimore,
MD 21201, USA.
Tel: 410 706 8181   Fax: 410 706 8184
lederer@umbi.umd.edu
www.umbi.umd.edu/~mbc/pages/lederer.htm

Specialty Keywords: Heart, Confocal Microscopy, Patch Clamp, Calcium.

Work in the lab focuses on Ca^{2+} signaling in cardiac and other living cells. By combining confocal, multiphoton or wide-field microscopy with whole cell patch clamp techniques, we have been able to investigate the effects of subcellular and intracellular Ca^{2+} concentration ($[Ca^{2+}]i$ on cellular function. Diverse additional tools are used as needed including flash photolysis of caged chemicals, multi-photon uncaging, single channel examination in planar lipid bilayers and by patch clamp, immuno-fluorescence imaging, use of cells from transgenic and gene knockout animals, and use of primary cultures and co-cultures. Much of the recent work focuses on "calcium sparks" and how the heart works in health and disease.

-Nelson, M.T., Cheng, H., Rubart, M., Santana, L.F., Bonev, A., Knot, H. & Lederer, W.J. (1995). Relaxation of arterial smooth muscle by calcium sparks. Science 270:633-637.

Date submitted: Editor Retained.

Thomas S. Lee, M.Sc.

International School of Photonics,
Cochin University of Science and Technology,
Kalamassery, Cochin, India 682 022,
India.
Tel: +91 48 457 5848   Fax: +91 48 457 6714
lee@cusat.ac.in
www.photonics.cusat.edu

Specialty Keywords: Sol-gel, pH Sensors, Fiber Optic Sensors.

I have carried out extensive research in the development of fiber optic sensors for chemical and physical applications. Chemical sensors include pH sensors based on dye impregnated sol-gel coatings. Also I have prepared bulk dye doped xerogels for quantum yield measurements, thermal lens spectroscopy and nonlinear applications in collaboration with other scientists.

-Thomas Lee S, B Aneeshkumar, P Radhakrishnan, C P G Vallabhan and V P N Nampoori, *A microbent fiber optic pH sensor,* Opt. Comm 205, 253 – 256 (2002).
-Thomas Lee S, Nibu A George, P Sureshkumar, P Radhakrishnan, C P G Vallabhan and V P N Nampoori, *Chemical sensing with microbent optical fiber*, Opt. Lett., 20, 1541-1543 (2001).

Date submitted: Editor Retained.

Frank Lehmann, Ph.D.

Dyomics GmbH.,
Winzerlaer Str. 2, Jena,
D-07745,
Germany.
Tel: +49 364 150 8200   Fax: +49 364 150 8201
f.lehmann@dyomics.com
www.dyomics.com

Specialty Keywords: Biolabels, Probes, Multicolour Assays, FRET, HTS.
My current research interest is focussed on fluorescent labels for biological targets. I am involved in the design and customizing of reactive fluorophores with respect to their (photo)physical properties. The chromophores are mostly based on polymethines with cumarin or benzopyrylium heterocycles allowing easily to generate emission in the red and NIR region.

-P. Czerney, F. Lehmann, M. Wenzel, V. Buschmann, A. Dietrich and G.J. Mohr (2001). Tailor-Made Dyes for Fluorescence Correlation Spectroscopy *Biol. Chem.* 382(3) 495-498.

-P. Czerney, M. Wenzel, F. Lehmann and B. Schweder (2003). Compound, in particular marker-dye, based on polymethines *EP1318177A2*.

Date submitted: 4th April 2006

Barry R. Lentz, Ph.D.

Program in Molecular & Cellular Biophysics,
University of North Carolina at CH,
Chapel Hill, NC 27599-7260,
USA.
Tel: 919 966 5384   Fax: 919 966 2852
uncbrl@med.unc.edu
hekto.med.unc.edu:8080/FACULTY/LENTZ/lab.html

Specialty Keywords: Membrane Probes, Phase Fluorescence, Fusion, Phosphatidylserine, Blood Coagulation.
Dr. Lentz's lab uses fluorescence spectroscopy to define lipid regulation of blood coagulation. Through the use of model membranes and soluble forms of phosphatidylserine (PS), Lentz has identified regulatory PS sites on several key human coagulation proteins, and shown that PS, which is exposed during platelet activation, is an allosteric regulator of these proteins and a second messenger in blood coagulation. Lentz's lab is also a leader in the application of fluorescence and other methods to studying the mechanism of fusion in model membranes induced to fuse by polyethylene glycol. Lentz uses fusion proteins reconstituted into model membranes to examine the role of fusion proteins in catalyzing fusion during viral infection and neurotransmitter release.

Date submitted: 19th April 2006

Yao-Qun Li, Ph.D.

Department of Chemistry,
Xiamen University,
Xiamen 361005,
China.
Tel: / Fax: +86 592 218 5875
yqlig@xmu.edu.cn

Specialty Keywords: Fluorescence, Multi-component Analysis.
The research fields include molecular fluorescence spectroscopy & its application in environmental & biological analysis, multi-component analysis, and surface analysis. Special interests have focused on the development, instrumentation & application of some fluorescence techniques, such as synchronous fluorescence spectroscopy, multi-dimensional fluorescence, derivative technique, reflection fluorescence and confocal microscopy.

-Rapid and simultaneous determination of coproporphyrin and protoporphyrin in feces by derivative matrix isopotential synchronous fluorescence spectrometry, with Dan-Li Lin, Li-Fang He, *Clin. Chem*, 2004, 50(10), 1797.

-Spectral fluctuation and heterogeneous distribution of porphine on the water surface, with Maxim. N. Slyadnev, Takanori Inoue, Akira Harata and Teiichiro Ogawa, *Langmuir*, 1999, 15(*9*), 3035.

Date submitted: 15th July 2005

Panagiotis Lianos, Ph.D.

Engineering Science Department,
University of Patras,
26500 Patras,
Greece.
Tel: 30 261 099 7587   Fax: 30 261 099 7803
lianos@upatras.gr

Specialty Keywords: Luminescence of Nanocomposite Organic-Inorganic Gels.
AIM 2004 = 20.9

Current research interests are related with the study of nanocomposite organic-inorganic materials. Deposition as thin films and calcination leads to the formation of semiconductor nanocrystallites useful for photocatalytic applications. Nanocomposite materials may be luminescent and they are studied by fluorescent probing. They are also studied as hosts of luminescent species and are used in various applications.

- P.Lianos, J.Fluorescence, 14(2004)11-15.
- V.Bekiari, K.Pagonis, G.Bokias and P.Lianos, Langmuir, 20(2004)7972-7975.

Date submitted: 4th April 2006

David M. J. Lilley, FRS.

CR-UK Nucleic Acid Structure Research Group,
MSI/WTB Complex, University of Dundee,
Dow Street, Dundee, DD1 5EH,
United Kingdom.
Tel: +44 138 234 4243   Fax: +44 138 234 5893
d.m.j.lilley@dundee.ac.uk
www.dundee.ac.uk/biocentre/nasg/

Specialty Keywords: Nucleic Acids, FRET, Single Molecules

Our interests are directed at the structure and folding of branched nucleic acids; the four-way junction in DNA, and a variety of structures (especially ribozymes) in RNA. Our main biophysical approach is fluorescence resonance energy transfer (FRET), in steady state, time-resolved and single-molecule modes.

-J. Liu, A.-C. Déclais, S. McKinney, T. Ha, D.G. Norman and D.M.J. Lilley Stereospecific effects determine the structure of a four-way DNA junction *Chem. & Biol.* 12, 217-228 (2005).
-M.K. Nahas, T.J. Wilson, S. Hohng, K. Jarvie, D.M.J. Lilley and T. Ha Observation of internal cleavage and ligation reactions of a ribozyme *Nature Struct. Molec. Biol.* 11, 1107-1113 (2004).

Date submitted: 11th October 2006

M. Pilar Lillo, Ph.D.

Instituto Química Física "Rocasolano",
C.S.I.C., Dep. Biofísica,
Serrano 119, 28006 Madrid,
Spain.
Tel: 34 91 561 9400   Fax: 34 91 564 2431
pilar.lillo@iqfr.csic.es

Specialty Keywords: Time-resolved Fluorescence, FRET, Biomolecular Interactions, Crowding.

Current interest: Structure and dynamics of macromolecular complexes in crowded and confined media: Design of time-resolved fluorescence anisotropy, FRET methodologies

-S.Zorrilla, G.Rivas, A.U.Acuña, M.P.Lillo (2004). Protein self-association in crowded protein solutions: a time-resolved fluorescence polarization study. *Protein Science 13, 2960-2969.*
-S. Zorrilla, M. Hink, A.J.W.G. Visser and M.P. Lillo (2006). Translational and rotational motions of proteins in a protein crowded environment. *Biophysical Chemistry* doi:10.1016/j.bpc.2006.09.003

Date submitted: Editor Retained.

Marcin Lipski, Ph.D.

Poznan University of Technology,
Institute of Chemistry & Technical Electrochemistry,
Piotrowo 3, 60 965 Poznań,
Poland.
Tel: +48 (61) 665 2068   Fax: +48 (61) 665 2571
mlipski@sol.put.poznan.pl
www.put.poznan.pl

Specialty Keywords: Photochemistry & Molecular Spectroscopy of Humic Acids & Precursors-Hydroxybenzotropolones.

Current Research Interests: Fluorescence of humic acids and unusual precursors - purpurogallin (2,3,4,6–tetrahydroxy–5H-benzocyclohepten–5–one, hydroxybenzotropolone) and its analogues formed from the polyphenols.

-M. Lipski (2002). Fluorescence emitted during the autooxidation of 2,3,4,6-tetrahydroxy-5H-benzocyclohepten-5-one, *Journal of Fluorescence,* 12(1), 83-86.
-M. Lipski, K. Gwozdzinski, J. Slawinski (2000). Free radical of the semiquinone type generated in the redox reaction of hydroxybenzotropolone, *Current Topics in Biophysics,* 24(2), 115-120.
-M. Lipski, J. Slawinski, D. Zych (1999). Changes in the luminescent properties of humic acids induced by UV-radiation, *Journal of Fluorescence,* 9(2), 133-138.

Date submitted: 24th May 2006

Burton J. Litman, Ph.D.

Section of Fluorescence Studies,
Laboratory of Membrane Biochemistry and Biophysics,
National Institute on Alcohol and Alcoholism,
National Institutes of Health, 12420 Parklawn Drive, Rm 114.
Rockville, Maryland, 20852 USA
Tel: 301 594 3608   Fax: 301 594 0035
litman@helix.nih.gov

Specialty Keywords: Membrane Structure, Fluorescent Probes, GPCR Signaling Systems.

Research interests focus on the effect of lipid composition on GPCR signaling, using the visual transduction system as a model. The role of polyunsatrurated phospholipids and cholesterol in modulating signaling and domain formation is investigated. Membrane phospholipid acyl chain packing and domain formation are monitored using various fluorescence techniques.

-Trans fatty acid derived phospholipids show increased membrane cholesterol and reduced receptor activation as compared to their cis analogs (2005) Niu SL, Mitchell DC, Litman BJ, BIOCHEMISTRY 44 (11): 4458-4465.
-Reduced G protein-Coupled Signaling Efficiency in Retinal Rod Outer Segments in Response to N-3 Fatty Acid Deficiency, (2004) Shui-Lin Niu, Drake C. Mitchell, Sun-Young Lim, Zhi-Ming Wen, Hee-Yong Kim, Norman Salem, Jr., and Burton J. Litman, J.Biol. Chem 279 31098-31104.

Date submitted: Editor Retained.

Garrick M. Little, Ph.D.

Li-Cor,
4308 Progressive Ave, Lincoln,
Nebraska, 68504,
USA.
Tel: (402) 467 0716   Fax: (402) 467 0819
glittle@licor.com
www.licor.com

Specialty Keywords: Protein Labeling, Western Blot Assay, DNA Labeling.

My research interests include the synthesis of Infra-red fluorescent dyes functionalized as the amidite, NHS ester etc. Labeling of biological molecules with fluorescent dyes. More generally Organic chemistry synthesis, synthesis of DNA.

Date submitted: 22nd July 2004

David Lloyd, Ph.D., D.Sc.

Microbiology (BIOSI1), Cardiff University,
P.O. Box 915, Cardiff,
CF10 3TL, Wales,
UK.
Tel: +44 292 087 4772   Fax: +44 292 087 4305
lloydd@cf.ac.uk
www.cf.ac.uk/biosi/research/micro/staff/dl.html

Specialty Keywords: Mitochondria, 2-Photon, Flow Cytometry.

Bioenergetics of lower eukaryotes especially yeasts and protists. Mitochondria and hydrogenosomes.Biological oscillations and clocks, especially ultradian timekeepers.Non-invasive monitoring of cell structure and function. Low oxygen measurements.

-Cycles of mitochondrial energisation driven by the ultradian clock in a culture of Saccharomyces cerevisiae .Microbiology 148,3715- 3724 (2002).

-The plasma membrane of microaerophilic proists;oxidative and nitrosative stress . Microbiology 150,1231-1236(2004).

Date submitted: 10th October 2006

Leslie M. Loew, Ph.D.

Center for Cell Analysis and Modeling,
University of Connecticut Health Center,
Farmington, CT 06030 1507,
USA.
Tel: 860 679 3568   Fax: 860 679 1039
les@volt.uchc.edu
www.ccam.uchc.edu/

Specialty Keywords: Non-linear Optical Microscopy, Dye
Synthesis, Cell Physiology.

We have a long-standing effort on the synthesis of voltage-sensitive dyes which has recently led us to develop dyes and optical systems for second harmonic imaging microscopy. We have also been developing a computational system called "Virtual Cell" for modeling and simulating cellular events based on microscope images. Our biological research focuses on mechanisms that spatially organize cellular signals.

-Jin, L., A. C. Millard, J. P. Wuskell, X. Dong, D. Wu, H. A. Clark and L. M. Loew. 2006. Characterization and Application of a New Optical Probe for Membrane Lipid Domains. Biophys. J., 90:2563-75.

-Yan, P., A.C. Millard, M. Wei, and L.M. Loew. 2006. Unique Contrast Patterns from Resonance-Enhanced Chiral SHG of Cell Membranes. J. Am. Chem. Soc. 128:11030-11031.

Date submitted: 6th July 2004

Piet H. M. Lommerse, Ph.D.

Dept. of Biophysics, Leiden University,
Niels Bohrweg 2, Leiden,
2333 CA,
The Netherlands.
Tel: +31 71 527 5946   Fax: +31 71 527 5819
lommerse@physics.leidenuniv.nl
www.biophys.leidenuniv.nl/Research/FvL

Specialty Keywords: Single-molecule Fluorescence
Microscopy.

In the last decade evidence has accumulated that small domains (50-700 nm diameter) are located in the plasma membrane. Using wide-field fluorescence microscopy with single-molecule sensitivity, the diffusion of individual membrane-anchored eYFP molecules is studied in live cells at the millisecond timescale [1], to reveal the intricate details of membrane organization and its role in signal transduction.

-P. H. .M. Lommerse, G. A. Blab, L. Cognet, G. S. Harms, B. E. Snaar-Jagalska, H. P. Spaink and T. Schmidt (2004). Single-molecule imaging of the H-Ras membrane-anchor reveals domains in the cytoplasmic leaflet of the cell membrane *Biophysical Journal* 86, 609-616.

Date submitted: 20th October 2006 **Fernando López Arbeloa, Ph.D.**

Department of Physical Chemistry,
University of the Basque Country UPV / EHU,
P.O. Box 644, Bilbao- 48080,
SPAIN.
Tel: +34 94 601 5971   Fax: + 34 94 601 35 00
fernando.lopezarbeloa@ehu.es
www.ehu.es/especmolecular/
Specialty Keywords: Photophysics, Laser Dyes, Dye/Clay Films.
AIM 2006 = 15.1

The photophysics of laser dyes and other fluorescent aromatic compounds are studied in a multitude of environments (liquid solutions, solid polymeric matrices, clay colloidal suspensions, layered solid thin films) in order to improve the photophysical and optical behavior of photonic materials. Development of a new methodology based on the fluorescence polarization to evaluate the orientation of fluorescent molecules adsorbed in rigid 2D layered materials.

-F. López Arbeloa et al. (2005). Structural, Photophysical and Lasing Properties of Pyrromethene Dyes. *Int. Rev. Phys. Chem.* 24, 339-374
-F. López Arbeloa, et al. (2006). New Fluorescent Polarization Method to Evaluate the Orientation of Adsorbed Molecules in Uniaxial 2D Layered Materials, *J. Photochem. Photobiol.* A181, 44-49

Date submitted: 20th October 2006 **Iñigo López Arbeloa, Ph.D.**

Department of Physical Chemistry,
University of the Basque Country UPV / EHU,
P.O. Box 644, Bilbao- 48080,
SPAIN.
Tel: +34 94 601 5972   Fax: + 34 94 601 3500
inigo.lopezarbeloa@ehu.es
www.ehu.es/especmolecular/
Specialty Keywords: Photophysics, Laser Dyes, Dye/Clay Films.
AIM 2005 = 14.7

The photophysics of laser dyes and other fluorescent aromatic compounds are studied in a multitude of environments (liquid solutions, solid polymeric matrices, clay colloidal suspensions, layered solid thin films) in order to improve the photophysical and optical behavior of photonic materials. Development of a new methodology based on the fluorescence polarization to evaluate the orientation of fluorescent molecules adsorbed in rigid 2D layered materials.

-I. López Arbeloa et al. (2004). Intramolecular Charge Transfer in Pyrromethene Laser Dyes: Photophysical Behaviour of PM650. *CehmPhysCehm.* 5, 1762-1771.
-I. López Arbeloa, et al. (2005). Orientation of Adsorbed Dyes in the Interlayer Space of Clay, *Chem. Mat..* 17, 4134-4141.

Lopez, A.
Losytskyy, M. Y.

Date submitted: Editor Retained.

André Lopez, Ph.D.

Institut de Pharmacologie et de Biologie Structurale du CNRS,
205 route de Narbonne,
Toulouse, 31400,
France.
Tel: (33) 56 117 5945   Fax: (33) 56 117 5994
andre.lopez@ipbs.fr
ipbs.fr

Specialty Keywords: Membrane Probes, Multichromophoric
Systems, Biomembranes.

Functional consequences of membrane composition and microcompartmentation in connection with the translational dynamics of lipids and proteins on the chain of signal transduction by G protein-coupled receptors. These studies are carried out on human receptors μ, CCR5, CXCR4 expressed in various cell types. Are investigated: (i) the influence of lipid environmental factors on receptor activity, (ii) the lateral dynamics and compartimentations of these membrane compounds using fluorescence techniques (FRAP, SPT), (iii) the structure *in situ* of these pluri-molecular systems by means of spectromicrofluorescence approaches (FRET, polarity probes).

Date submitted: 18th October 2006   **Mykhaylo Yu. Losytskyy, Ph.D.**

Kyiv Taras Shevchenko National Uni., Physics Department,
Acad. Glushkova Ave. 6, 03680 Kyiv, Ukraine,
Inst. of Molecular Biology and Genetics of NAS of Ukraine,
Zabolotnogo Str. 150, Kyiv 03143, Ukraine.
Tel: +38 044 526 3449   Fax: +38 044 522 2458
m_losytskyy@svitonline.com   &   mlosytskyy@gmail.com
www.yarmoluk.org.ua

Specialty Keywords: Dye-DNA Interaction, J-aggregate, Dye
Photophysics.

M. Losytskyy is working at the Department of Combinatorial Chemistry of IMBG under the guiding of Dr. S.Yarmoluk. The work of M. Losytskyy is aimed on designing of fluorescent probes for nucleic acid and protein detection. His studies are also devoted to DNA-dye binding equilibrium; H- and J-aggregates of cyanine dyes; and photophysics of the dye molecules.

-M.Yu. Losytskyy, K.D. Volkova, V.B. Kovalska, et al. (2005) J. Fluoresc., *15*, 849-857.

-M.Yu. Losytskyy, V.M. Yashchuk, S.S. Lukashov, S.M. Yarmoluk (2002) *J. Fluoresc. 12*, 109-112.

Date submitted: 29th July 2004

Luís M. S. Loura, Ph.D.

Centro de Química-Física Molecular, IST,
Av. Rovisco Pais,
1049-001 Lisbon,
Portugal.
Tel: +351 21 841 9219   Fax: +351 21 846 4455
pclloura@alfa.ist.utl.pt

Specialty Keywords: FRET, Lipid Domains, Lipid-protein Interaction.
AIM 2003 = 12.2

Study of membrane heterogeneities (domains/rafts) using photophysical methodologies. Derivation of kinetic models for FRET in restricted geometries. Development of software for global analysis of fluorescence decays. Topology and dynamics of protein/peptide interaction with model systems of membranes. Characterization of DNA/cationic lipid complexes.

-F. Fernandes, L. M. S. Loura, M. Prieto, R. Koehorst, R. Spruijt, and M. A. Hemminga. (2003) Dependence of M13 major coat protein oligomerization and lateral segregation on bilayer composition. *Biophys. J.* 85 (4), 2430-2441.

-C. Madeira, L. M. S. Loura, M. R. Aires-Barros, A. Fedorov, M. Prieto. (2003) Characterization of DNA/lipid complexes by fluorescence resonance energy transfer. *Biophys. J.* 85 (5), 3106-3119.

Date submitted: 24th October 2006

Parkash L. Mandhan, Ph.D.

Department of Paediatric Surgery,
Waikato Hospital,
Hamilton, 3210,
New Zealand.
Tel: +64 7 855 1918
kidscisurg@mac.com

Specialty Keywords: Developmental molecular biology.
AIM 2006 = 7.4

I have special interest in developmental molecular biology and my current research project involves role of shh and downstream genes in the development of hindgut in ETU-exposed fetal rats.

-Mandhan P,Qi BQ,Keenan JI, et al.(2006). Counterstaining improves visualization of the myenteric plexus in immunolabelled whole-mount preparations. *J Fluoresc* 16(5):655-658

-Mandhan P,Qi BQ,Beasley SW, et al.(2006). Sonic hedgehog expression in the of hindgut in ETU-exposed fetal rats *Pediatr Surg Int* 22 (1):31-36

Date submitted: 6th July 2006

Jeffrey G. Manni, M.Sc.

JGM Associates Inc.,
25 Mall Road, Suite 300,
Burlington, MA 01803,
USA.
Tel: 781 272 6692 Fax: 781 221 7154
jgmanni@jgma-inc.com
www.jgma-inc.com

Specialty Keywords: Lasers, Tunable, Solid-state.

Mr. Manni designs, develops, and manufactures tunable solid-state lasers. He is currently developing ultra-compact solid-state laser sources that can be tuned throughout the visible and near-IR wavelength regions. One tunable laser provides nanosecond pulsed emission, and the other generates tunable picosecond emission.

These lasers will facilitate the development of new laser-induced fluorescence methods and related instrumentation, especially ones that employ highly wavelength-multiplexed techniques.

Date submitted: Editor Retained.

Mark Maroncelli, Ph.D.

Department of Chemistry, Penn State University,
152 Davey Laboratory, University Park,
PA, 16802,
USA.
Tel: 814 863 5319
mpm@chem.psu.edu
maroncelli.chem.psu.edu

Specialty Keywords: Time-resolved Fluorescence, Ultrafast Spectroscopy, Solution Dynamics.

We use steady-state and ultrafast fluorescence spectroscopy and computer simulations to explore solvation and its influence over chemical processes in liquid solvents and supercritical fluids.

-J. Lewis, R. Biswas, A. Robinson, and M. Maroncelli (2001)., Local Density Augmentation in Supercritical Fluids: Electronic Shifts of Anthracene Derivatives *J. Phys. Chem. B* 105, 3306.

-M. L. Horng, J. A. Gardecki, A. Papazyan, and M. Maroncelli (1995)., Sub-Picosecond Measurements of Polar Solvation Dynamics: Coumarin 153 Revisited *J. Phys. Chem.* 99, 17311.

Date submitted: 17th September 2004 **José M. G. Martinho, Ph.D.**

Centro de Química-Física Molecular,
Instituto Superior Técnico,
1049-001 Lisboa,
Portugal.
Tel: +351 21 841 9250 Fax +351 21 846 4455
jgmartinho@ist.utl.pt

Specialty Keywords: Photophysical Kinetics, Resonance Energy Transfer, Polymers, Colloids.

Current interests: Conformation and dynamics of proteins and oligonucleotides adsorbed onto latex particles. Photophysical kinetics (early work included the study of transient effects in pyrene monomer-excimer kinetics). Radiative transport of electronic energy. Conformation and aggregation of polymers in solution. Polymer interfaces.

-T. J. V. Prazeres, A. Fedorov, J. M. G. Martinho (2004). Dynamics of Oligonucleotides Adsorbed on Thermosensitive Core-Shell Latex Particles, *J. Phys. Chem. B* 108 9032-9041.
-S. Piçarra, J. Duhamel, A. Fedorov, J. M. G. Martinho (2004). Coil-Globule Transition of Pyrene-Labeled Polystyrene in Cyclohexane: Determination of Polymer Chain Radii by Fluorescence, *J. Phys. Chem. B* 108 12009-12015.

Date submitted: Editor Retained. **Masayuki Masuko, Ph.D.**

Hamamatsu Photonics K. K., Tsukuba Research Laboratory,
5-9-2 Tokodai, Tsukuba,
300-2635,
Japan.
Tel: +81 29 847 5161   Fax: +81 29 847 5266
masuko@hpk.trc-net.co.jp
www.hpk.co.jp/

Specialty Keywords: Nucleic Acids, Excimer Fluorescence, Photon Counting.

I am interested in the application of aromatic hydrocarbon dyes to the detection of biological substances such as nucleic acids, and the development of instruments useful to their measurements.

-M. Masuko, H. Ohtani, K. Ebata and A. Shimadzu (1998) Optimization of excimer-forming two-probe nucleic acid hybridization method with pyrene as a fluorophore *Nucleic Acids Res.* 26 (23), 5409-5416.
-M. Masuko, S. Ohuchi, K. Sode, H. Ohtani and A. Shimadzu (2000) Fluorescence resonance energy transfer from pyrene to perylene labels for nucleic acid hybridization assays under homogeneous solution conditions *Nucleic Acids Res.* 28 (8), e34.

Mateo, C. R.
Matkó, J.

Date submitted: Editor Retained.

C. Reyes Mateo, Ph.D.

Instituto de Biología Molecular y Celular,
Universidad Miguel,
Ada. Ferrocarril s/n,
03202-Elche (Alicante), Spain.
Tel: +34 96 665 8469   Fax: +34 96 665 8758
rmateo@umh.es

Specialty Keywords: Lipid Membranes, Biosensors, Time-resolved Fluorescence Depolarization.

- Structure and dynamics of lipid membranes from time-resolved fluorescence depolarisation.
- Interaction, location and dynamics of proteins, peptides and small bioactive molecules in phospholipid model membranes.
Encapsulation of macromolecules in sol-gel matrices.
- Design and characterization of fluorescent biosensors with application in clinical diagnosis.
J.A. Poveda, M. Prieto, J.A. Encinar, J.M. González-Ros and C. R. Mateo (2003). Intrinsic tyrosine fluorescence as a tool to study the interaction of the shaker B "ball" peptide with anionic membranes. Biochemistry 42, 7124-7132.

Date submitted: 20th October 2006

János Matkó, Ph.D., D.Sc.

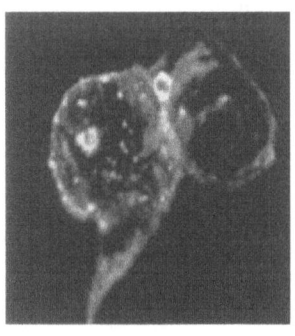

Department of Immunology,
Eötvös Lorand University,
Pázmány Péter sétány 1/C,
Budapest, H-1117, Hungary.
Tel: 36 1 381 2175   Fax: 36 1 381 2176
matko@elte.hu

Specialty Keywords: Flow Cytometry, Fluorescence Spectroscopy and Cell Imaging, Cell FRET.
AIM 2006 = 17.0

Research is conducted on cells of the immune system using (and developing) fluorescence cell analytical technologies in the following fields: characterization/functional role of plasma membrane lipid microdomains (rafts) and protein clustering; receptor-mediated signal transduction/signal networks in lymphocyte activation and apoptosis; structure and function of immunological synapses.

-Detre C, Kiss E, Varga Z, Ludányi K, Pászty K, Enyedi Á, Kövesdi D, Panyi G, Rajnavölgyi É, Matkó J: Death or survival: Membrane ceramide controls the fate and activation of antigen-specific T cells depending on signal strength and duration. Cellular Signalling 2006. 18:294-306.

Date submitted: 9th May 2006

James R. Mattheis, Ph.D.

Fluorescence Division,
HORIBA Jobin Yvon,
3880 Park Avenue,
Edison, NJ 08820-3012, USA.
Tel: 732 494 8660 Ext: 122   Fax: 732 549 5157
Jim_Mattheis@jyhoriba.com
www.jobinyvon.com

Specialty Keywords: Fluorescence Imaging, Frequency-domain lifetime, TCSPC.

Dr. Mattheis manages a team of scientists providing fluorescence applications support, training and new methods development for users of SPEX spectrofluorometers. Support is provided for all users interested in applying high sensitivity photon-counting, steady-state fluorescence spectroscopy, fluorescence microscopy and picosecond time-resolved techniques to their own research projects. We support both frequency-domain and time correlated single photon counting instrumentation and are continuing the development of new and novel light sources.

Date submitted: 24th April 2006

Yvette M. Mattley, Ph.D.

Ocean Optics, Biophotonics Division,
830 Douglas Avenue,
Dunedin, FL 34698,
USA.
Tel: 727 733 2447   Fax: 727 733 3962
Yvette.Mattley@OceanOptics.com

Specialty Keywords: Fluorescence, Photoluminescence, Biophotonics, Biochemistry, Fluorescence Spectroscopy.

Dr. Mattley heads the Ocean Optics Biophotonics Division and focuses on the development of fluorescence-based instrumentation and applications. Her projects include the EDS2000, a spectrophotometric system for the rapid detection of bacterial endospores based on the enhancement of terbium photoluminescence occurring in the presence of a major spore component. Dr. Mattley's application research includes fluorescent tagging using quantum dots and organic fluorophores, chlorophyll analysis, anthrax detection and FRET analysis of PCR products.

-Mattley, Y. Fluorescence Spectroscopy Goes Miniature. *Laboratory Equipment*, Apr. 2003, pp. 14-15; Mattley, Y., Bacon, T., DeFrece R. Miniature Spectroscopic Instrumentation:Applications to Biology and Chemistry, *Review of Scientific Instruments*, Vol. 75, # 1, Jan. 2004, pp. 1-16.

Mátyus, L.
V. M. Mazhul,

Date submitted: 6th April, 2006

László Mátyus, M.D., Ph.D., D.Sc.

Department of Biophysics and Cell Biology,
University of Debrecen,
Nagyerdei krt 98, Debrecen,
H-4012, Hungary.
Tel: / Fax: +36 5 241 2623
lmatyus@jaguar.unideb.hu

Specialty Keywords: Fluorescence Resonance Energy Transfer,
Flow Cytometry.
AIM 2004 = 21.3

Current research interests: study the distribution and conformation of cell surface receptors using fluorescence techniques, such as flow cytometric energy transfer measurements or different microscopies.

-L. Mátyus, J. Szöllősi, A. Jenei (2006). Steady-state fluorescence quenching applications for studying protein structure and dynamics *J. Photochem. Photobiol. B: Biol.* 83, 223-236.
-G. Szentesi, G. Horváth, I. Bori, G. Vámosi, J. Szöllősi, R. Gáspár, S. Damjanovich, A. Jenei and L. Mátyus (2004). Computer program for determining fluorescence resonance energy transfer efficiency from flow cytometric data on a cell-by-cell basis *Comp. Meth. Prog. Biomed.* 75, 201-211.

Date submitted: 5th July 2005

Vladimir M. Mazhul, Ph.D.

Laboratory of Proteomics, Institute of Biophysics & Cellular
Engineering of National Academy of Sciences of Belarus,
Akademicheskaya 27, Minsk,
Belarus, 220072.
Tel: 375 17 284 2358   Fax: 375 17 284 2359
mazhul@biobel.bas-net.by

Specialty Keywords: Room Temperature Phosphorescence.

Corresponding Member of National Academy of Sciences of Belarus. Specialist in the fields of studying proteins and lipid peroxidation (LPO) products by fluorescence and room temperature phosphorescence techniques. The systematic investigations of millisecond internal dynamics of proteins in solution and composition of cell membrane by room temperature tryptophan phosphorescence technique had been carried out. By room temperature phosphorescence method the heterogeneity of LPO products accumulation in bulk and annular lipids of the cellular membrane has been shown.

-V.M. Mazhul, E.M. Zaitseva, M.M. Shavlovsky, O.V. Stepanenko, I.M. Kuznetsova, K.K. Turoverov. // Biochemistry, 2003, V. 42, P. 13551-13557.
-V. Mazhul, T. Chernovets, E. Zaitseva, D. Shcharbin. // Cell Biology International, 2003, V. 27, P. 571-578.

Date submitted: 11th September 2006

Alberto Mazzini, Ph.D.

Department of Physics,
University of Parma,
Via Usberti 7A,
Parma, 43100, Italy.
Tel: +39 052 190 6229   Fax: +39 052 190 5223
mazzini@fis.unipr.it

Specialty Keywords: Protein Folding, Binding Analysis of Probes to Proteins, Time Correlated Single Photon Counting.

My present research interest is to study denaturation and renaturation mechanisms of proteins. Unfolding is induced by chemical denaturants, temperature and pH, refolding is obtained by recovery the native experimental conditions. Fluorescence measurements on intrinsic and extrinsic probes are made both by stationary and time resolved techniques (TCSPC). In the case of simple monomeric or dimeric proteins such as odorant binding proteins (OBP), the thermodynamic and kinetic analysis allows to elucidate the unfolding/refolding mechanism.

-M. Parisi, A. Mazzini, R.T. Sorbi, R. Ramoni, S. Grolli and R. Favilla
Biochim. Biophys. Acta 1750, 30 (2005).
-F. Tian, K. Johnson, A.E. Lesar, H. Moseley, J. Ferguson, I.D.W. Samuel, A. Mazzini and L. Brancaleon Biochim. Biophys. Acta 1760, 38 (2006).

Date submitted: 6th April 2006

Igor L. Medintz, Ph.D.

Center for Bio/Molecular Science and Engineering, Code 6900,
U.S. Naval Research Laboratory,
4555 Overlook Avenue, S.W.,
Washington D.C., 20375, USA.
Tel: 202 404 6046   Fax: 202 767 9594
Imedintz@cbmse.nrl.navy.mil
www.nrl.navy.mil/   &   nrlbio.nrl.navy.mil/

Specialty Keywords: Quantum Dots, FRET, Biosensors.
AIM 2005 = 40.9

The current research is two-fold: adapting and utilizing photoluminescent semiconductor nanoparticles for a variety of biosensing assays and creating reagentless biosensors utilizing multifunctional modular protein-DNA structures. Both projects exploit FRET as the mechanism of signal transduction.

-I.L. Medintz, H.T. Uyeda, E.R. Goldman and H. Mattoussi (2005). Quantum dot bioconjugates for imaging, labeling and sensing. *Nature Materials* 4,435-446.
-I.L. Medintz, J.H. Konnert, A.R. Clapp, I. Stanish, M.E. Twigg, H. Mattoussi, J.M. Mauro and J.R. Deschamps (2004). A fluorescence resonance energy transfer derived structure of a quantum dot-protein bioconjugate nanoassembly. *P.N.A.S.U.S.A.* 101,9612-9617.

Melnyk, J. M.
Mely, Y.

Date submitted: 11th October 2006

James M. Melnyk.

Boston Electronics Corporation,
91 Boylston Street, Brookline,
MA 02445,
USA.
Tel: 800 347 5445 / 617 566 3821   Fax: 617 731 0935
jmm@boselec.com
www.boselec.com

Specialty Keywords: Photodetection, APDs, SPADs.

James M Melnyk is an Applications Specialist for photodetectors at Boston Electronics Corporation, North American agents for Becker & Hickl GmbH of Berlin, Germany; for Edinburgh Instruments Ltd of Edinburgh, Scotland; and for id Quantique of Geneva, Switzerland.

Specialist in photodetection for photon counting and analytical instruments.

Date submitted: 11th October 2006

Yves Mely, Ph.D.

Université Louis Pasteur, UMR 7034 CNRS,
Faculté de Pharmacie, 74 route du Rhin,
67401 Illkirch,
France.
Tel: 33 (0)39 024 4263   Fax: 33 (0)39 024 4213
mely@pharma.u-strasbg.fr
umr7034.u-strasbg.fr/
Specialty Keywords: Time-resolved Fluorescence, Fluorescence Correlation Spectroscopy, Interactions, Two colour probes.
AIM 2004 = 35.2

The research of my team is mainly focused on the investigation by fluorescence techniques of the interaction of proteins (mainly HIV nucleocapsid protein) with ligands in solution and cells. We also investigate the physico-chemical properties and intracellular fate of complexes of DNA with nonviral vectors. Finally, we develop fluorescent probes (mainly two color probes) as well as a two photon platform that combines FCS, time-resolved fluorescence and imaging.

-N. Ben Gaied et al. 8-vinyl-deoxyadenosine, an alternative fluorescent nucleoside analogue to 2'-deoxyribosyl-2-aminopurine with improved properties. Nucleic Acids Research, 2005, 33, 1031-1039.

-H. Beltz et al. Structural determinants of HIV-1 nucleocapsid protein for cTAR DNA binding and destabilization. J. Mol. Biol., 2005, 348, 1113-1126.

Date submitted: 16th March 2004

Francisco Mendicuti, Ph.D.

Química Física, Univesidad de Alcalá,
Ctra Madrid-Barcelona Km 33.6,
28871 Alcalá de Henares, Madrid,
Spain.
Tel: 34 91 885 4672   Fax: 34 91 885 4763
francisco.mendicuti@uah.es

Specialty Keywords: Excimers, Energy Transfer, Polymers, Inclusion Complexes, Molecular Mechanics / Dynamics.

We use steady state and time-resolved fluorescence techniques, as well as various theoretical methods for the study of some conformational properties in polymer systems and the inclusion processes of small molecules and polymers with cyclodextrins. Comparison of the theoretical and experimental results allow us to learn more about the conformations and dynamics of polymeric systems and the driving forces and thermodynamics accompanying complexation processes.

-Gallego, J, Mendicuti, F., Mattice, W.L. *J. Polym. Sci. Polym.Phys. E.* 2003, 41, 1615.

-Dimarino, A., Mendicuti, F. *Appl. Spectrosc..* 2004, 58(7), 823.

Date submitted: Editor Retained.

Svetlana B. Meshkova, D.Sc., Ph.D.

Department of Analytical Chemistry and Physico-Chemistry of Coordination Compounds, A.V. Bogatsky Physico-Chemical Institute of National Academy of Sciences of Ukraine, National Academy of Sciences of Ukraine.
86, Lustdorfskaya doroga, 65080, Odessa, Ukraine.
Tel: +38(0482) 652 042   Fax: +38(0482) 652 012
physchem@paco.net

Specialty Keywords: Fluorescence, Energy Transfer, Lanthanide Complexesl.

Current Research Interests: Design and investigation of photochemical properties of lanthanide complexes in solution and solid matrix. Investigation of connection between the composition, stability and optical properties of complexes and characteristics of lanthanide ions and ligands. Study of new means for elimination of intra- and intermolecular energy losses and its realization in luminescent analysis.

-S.B. Meshkova (2000). The Dependence of the Luminescence intensity of Lanthanide Complexes with β-Diketones on the Ligand Form: J. of Fluorescence, 10(4), 333-337.

-S.B. Meshkova, Z.M. Topilova, D.V. Bolshoy, S.V. Beltyukova, M.P. Tsvirko and V.Ya. Venchikov (1999). Quantum Efficiency of the Luminescence of Ytterbium (III) β-Diketonates: Acta Phys. Polonica A, 95(6), 983-990.

Minet, O.
Mirochnik, A. G.

Date submitted: 5th April 2006

Olaf Minet, Ph.D.

Charité – Universitätsmedizin Berlin,
Campus Benjamin Franklin,
Biomedizinische Technik und Physik,
Fabeckstr. 60-62, 14195 Berlin, Germany.
Tel: 49 308 449 2311   Fax: 49 308 445 4377
olaf.minet@charite.de
www.fu-berlin.de

Specialty Keywords: Fluorescence Microscopy, Optical Biopsy, Optical Molecular Imaging, Quantum Dots, Image Processing.

Current Research Interests: My research is focused on advancing fluorescence applications in medicine. This involves native autofluorescence compounds like NADH in Optical Biopsy, synthetic markers in Optical Molecular Imaging, multi-color Quantum dots and fluorescence cell monitoring as well. Of special interest are investigations in the field of image processing, i.e. for eliminating the effects of tissue optics like absorption and scattering on the fluorescence signal by deconvolution.

-Minet O, Zabarylo U, Beuthan J: Deconvolution of laser based images for monitoring rheumatoid arthritis. Las. Phys. Lett. 2 (2005) 556-565

Date submitted: 20th October 2006

Anatolii G. Mirochnik, Ph.D.

Far-Eastern Branch of the Russian Academy of Sciences,
Institute of Chemistry,
159 prosp.100-letiya, Vladivostoka,
690022, Russia.
Tel: 42 3 231 0466   Fax: 42 3 231 1889
mirochnik@ich.dvo.ru

Specialty Keywords: Fluorescence, Structure, Photochemistry.

Dr. Mirochnik's research interests include: Design and investigation of fluorescence and photochemical properties of lanthanide and p-elements (boron, s^2 - ions) complexes. Study of photochemical reaction mechanisms, ascertainment of correlations between spectroscopic parameters and molecular structure.

-Mirochnik A.G., Fedorenko E.V., Kuryavyi V.G., Bukvetskii B.V., Karasev V.E. Luminescence and reversible luminescence thermochromism of bulk and microcrystals of dibenzoylmethanato-boron difluoride 2006, Journal of Fluorescence, 16 (3) , 279-286.

-Mirochnik A.G., Bukvetskii B.V.,Zhikhareva P.A., Polyakova N.V., Karasev V.E. Crystal Structure and Triboluminescence of the [Tb(BTFA)$_2$(NO$_3$)(TPPO)$_2$], 2006, Russian Journal of Inorganic Chemistry, 51(5) , 737-742.

Date submitted: 13th April 2006

Hirdyesh Mishra, Ph.D.

Molecular Biophysics Unit,
Indian Institute of Science,
Bangalore – 560 012,
India.
Tel: +91 802 293 2337   Fax: +91 802 360 0683
hirdyesh@mbu.iisc.ernet.in   &   hirdyesh@yahoo.com
Specialty Keywords: Time domain ultra fast fluorescence
spectroscopy / microscopy, Laser and Molecular spectroscopy,
Electronics and Electronic structure theoretical computation.
AIM 2006 = 6.3

Research Interest: My basic research interest is study of electronic structure and dynamics of photo-induced excited state proton, electron and energy transfer in hydrogen bonded molecules of interest in organized assemblies and nano-structured materials. Recently I also started to understand preparation and processing of lasing, photonic, nano-biophotonic materials and optical twizers. Besides this I am also interested to design and fabrication of instruments and programming for computation.

-Photo-induced Relaxation and proton transfer in some Hydroxy Naphthoic Acids in Polymers, H. Mishra, J. Phys. Chem. B. 110 (19) (2006) 9387 -9396.

-Edge Excitation Red Shift and Energy Migration in Quinine sulphate Dication, H. Mishra*, D. Pant, T. C. Pant, B. Tripathi; J. Photochem Photobiol: A 177 (2006) 197-204.

Date submitted: Editor Retained.

Tom Misteli, Ph.D.

National Cancer Institute, NIH,
41 Library Drive, Bldg. 41, B610,
Bethesda, MD 20892,
USA.
Tel: 301 402 3959   Fax: 301 496 4951
mistelit@mail.nih.gov
rex.nci.nih.gov/RESEARCH/basic/lrbge/cbge.html

Specialty Keywords: Living Cells, Photobleaching, Modelling.

My laboratory uses photobleaching, in situ hybridization and FCS methods to study nuclear architecture and genome expression in vivo. We make extensive use of kinetic modeling methods to analyze in vivo microscopy data.

-Phair R.B and T. Misteli, High mobility of proteins in the mammalian cell nucleus. Nature, 404, 604-609 (2000).
-Phair R.B. and T. Misteli, Kinetic modeling approaches to in vivo microscopy, Nature Rev. Mol. Cell Biol., 2, 898-907 (2001).

Date submitted: 7th August 2004

Ihab Kamal Mohamed, Ph.D.

Zoology Dept., Faculty of Science,
Ain-Shams Universcity, Cairo, Egypt.
& Cell Biology (LS. Plattner) Biology Dept.
Konstanz Uni., 78464 Konstanz, Germany.
Tel: +202 639 0470 Mobile: +20 10 663 5410
ihabkmohamed@yahoo.com
ihab.mohamed@uni-konstanz.de
www.ub.uni-konstanz.de/kops/volltexte/2002/760
Specialty Keywords: Calcium, Fluorochrome Analysis,
SOC Mechanism, Exocytosis, Paramecium, Secretion.

I use microinjection of different fluorochromes into living cells to detect Ca^{2+} mobilization across the cell membranes, in its stores and in mitochondria too. This is proceeded under CLSM or 2λ inverted microscope for time-resolved fluorescence imaging and by help of sophisticated computerized process. It is amazing to use the fluorochromes for localization of different cellular organelles, especially in living cells. I also use GFP application for localization and functional analysis of different proteins in living cells. I'll be glad for future scientific cooporation.

-Mohamed et al. 2003. Refilling of cortical calcium stores in Paramecium cells: in situ analysis in correlation with store-operated calcium influx. Journal of cell calcium, 34, pp. 87-96.

Date submitted:11th October 2006

Gerhard J. Mohr, Ph.D.

Institute of Physical Chemistry,
Friedrich-Schiller University,
Lessingstrasse 10, D-07743 Jena,
Germany.
Tel: +49 3641 94 8368   Fax: +49 3641 94 8302
gerhard.mohr@uni-jena.de
www.uni-jena.de/~c1moge/

Specialty Keywords: Luminescent Sensors, Reactands,
Sensor Nanoparticles.

Our research is dedicated to the development of new functional dyes and the investigation of their sensing properties in polymer layers. Then, they are adapted to miniaturized optics components for the detection of gaseous and dissolved analytes relevant in environmental, medical and biotechnical areas. Furthermore, we develop fluorescent sensor nanoparticles for monitoring ions and biomolecules in cells, and we adapt dyes for molecular imprinting.

-G. J. Mohr, Chromo- and fluororeactands: Indicators for detection of neutral analytes by using reversible covalent-bond chemistry. (2004). *Chemistry, A European Journal*, 10, 1083-1090.
-G. J. Mohr, Covalent bond formation as an analytical tool to optically detect neutral and anionic analytes. (2005). *Sensors and Actuators B*, 107, 2-13.

Date submitted: 30th June 2005

Laurent Monat.

idQuantique SA,
Chemin de la Marbrerie 3,
1227 Carouge-Geneva,
Switzerland.
Tel: +41 22 301 8376   Fax: +41 22 301 8379
laurent.monat@idquantique.com
www.idquantique.com

Specialty Keywords: Single photon Avalanche Diodes, Time-correlated Single photon Counting, Quenching Circuit, Pulser.

Laurent Monat's current research interests include: detectors and detector arrays fabricated in CMOS technologies, low timing jitter single photon avalanche diodes for short decay time measurements, electronic circuits for Geiger mode avalanche diodes, design of imaging devices for applications in medicine and biology.

Date submitted: 13th October 2006

Alexander Moroz, Ph.D.

Wave-scattering.com,
Gitschiner Strasse 106,
D-10969 Berlin,
Germany.
Tel: +49 302 590 1638   Fax: +49 302 590 1609
wavescattering@yahoo.com   &   moroz@phys.uu.nl
www.wave-scattering.com
Specialty Keywords: nanophotonics, photonic crystals, scattering.
AIM 2005 = 4.5

Dr. Moroz's current interests include: Various classical and quantum aspects of interaction of light with nanoparticles, nanocavities, and photonic crystal arrays, including elastic and inelastic scattering, fluorescence, decay rates, Rabi splitting, Förster energy transfer.

-A. Moroz, Chem. Phys. 117(1), 1-15 (2005)

-A. Moroz, Annals of Physics (NY) 315, 352-418 (2005)

Date submitted: 2nd August 2004   **Larry E. Morrison, Ph.D.**

Research and Development, Vysis / Abbott,
3100 Woodcreek Drive, Downers Grove,
Illinois, 60515,
USA.
Tel: 630 271 7136   Fax: 630 271 7128
lmorrison@vysis.com
www.vysis.com
Specialty Keywords: Fluorescence, In Situ Hybridization,
Energy Transfer Assays, DNA Labeling, Cancer Diagnostics.
AIM 2002 = 25.1

Current Research Interests: Developing diagnostic, prognostic, and predictive assays for human cancers employing both fluorescence *in situ* hybridization and PCR-based assays. This has included developing multi-target *in situ* hybridization technology using many fluorescent labels simultaneously, combinatorially, or ratiometrically. An early and continuing interest is homogeneous fluorescence detection systems, especially as applied to detecting PCR products.

-Morrison (2003) Fluorescence in nucleic acid hybridization assays, in J Lakowicz (Ed.) *Topics in Fluorescence Spectroscopy*, Vol 7. Kluwer, New York pp 69-97.
-Jacobson *et al.* (2004) Gene Copy Mapping of the *ERBB2/TOP2A* Region in Breast Cancer. *Genes, Chromosomes, and Cancer* 40, 19-31.

Date submitted: 24th June 2005   **Francis Müller, Ph.D.**

Pharma Research, F.Hoffmann-La Roche Ltd.,
Grenzacherstrasse 124,
CH-4070 Basel,
Switzerland.
Tel: +41 (0)61 688 6430   Fax: +41 (0)61 688 7408
francis.mueller@roche.com

Specialty Keywords: Protein Characterization, Binding Affinities, Spectroscopy.

Biomolecular structure: Proteins dynamics, mobility of tryptophanes for structural studies. Characterization of lead structures for protein binding. Hits validation from high throughput screening and biological assays. Support in fine tuning of potential ligands with quantitative measurement of affinities by fluorescence titration.

-A. Ruf, F. Müller, B. d'Arcy, M. Stihle, E. Kusznir, C. Handschin, O.H. Morand, R. Thoma (2004). The monotopic membrane protein human oxidosqualene cyclase is active as monomer. *BBRC* 315, 247-254.

Date submitted: 15th September 2005   **Mary-Ann Mycek, Ph.D.**

College of Engineering, Dept. of Biomedical Engineering,
Applied Physics Program, Comprehensive Cancer Center,
University of Michigan, 2200 Bonisteel Blvd.,
Ann Arbor, MI 48109-2099, USA.
Tel: 734 647 1361   Fax: 734 936 1905
mycek@umich.edu
www-personal.engin.umich.edu/~mycek/

Specialty Keywords: Biomedical Optics, Optical Diagnostics.

In my laboratory, we develop and apply advanced methods of optical and laser sciences to non-invasively probe complex living systems, with a current emphasis on sensing and quantifying physiological status. Research projects include optically sensing premalignant transformation in epithelial cells and tissues, metabolic activity in tissue engineered constructs, and hemodynamics in the brain. My translational research program in biomedical optics incorporates projects in basic (pre-clinical), applied (clinical), and computational science.

-M.-A. Mycek and B.W. Pogue, Eds.: Handbook of Biomedical Fluorescence, Biomedical Optics Series, Marcel-Dekker, Inc., New York (2003).

Date submitted: 22nd October 2006   **Stefan Nagl, Dipl. Chem.**

Institute of Analytical Chemistry, Chemo- and Biosensors,
University of Regensburg,
Universitätstr. 31, 93053 Regensburg,
Germany.
Phone: +49 (0) 941 943 4793   Fax: +49 (0) 941 943 4064
stefan.nagl@chemie.uni-regensburg.de
www-analytik.chemie.uni-regensburg.de
Specialty Keywords: Fluorescence Lifetime Imaging,
Bioassays, Nanoparticles.
AIM 2006 = 2.7

I am a graduate student in the group of Prof. Wolfbeis and work on time-resolved and fluorescence lifetime-based macro- and microscopic (FLIM) CCD-based imaging methods and techniques mainly for bioanalytical applications. These include optical oxygen and temperature sensors, DNA and protein microarrays and fluorescent and phosphorescent dye-doped polymer nanoparticles for protein analysis and as tracers in animal species.

-Nagl S, Schäferling M, Wolfbeis OS (2005) Fluorescence Analysis in Microarray Technology. Microchim. Acta 151(1-2):1-21

-Borisov SM et al. (2006) In: Berberan-Santos MN (ed) The Fluorescence of Nanostructured Systems. Springer, Vienna.

Date submitted: 11th May 2006

Ramaier Narayanaswamy, Ph.D., D.Sc.

School of Chemical Engineering & Analytical Science,
The University of Manchester, P.O. Box 88,
Sackville Street, Manchester M60 1QD,
United Kingdom.
Tel: +44 (0) 161 306 4891   Fax: +44 (0) 161 306 4911
ramaier.narayanaswamy@manchester.ac.uk

Specialty Keywords: Optical Sensors, Biosensors,
Instrumentation.

The main research involves the fundamentals and applications of molecular spectroscopy particularly, the development of optical chemical sensors and biosensors for various types of applications (environmental, industrial process control and biomedical). Analyte sensitive reagent matrices are interrogated using optical fibers in the sensor designs and the optical property measured are absorbance, reflectance and fluorescence. Signals are analyzed using signal processing techniques including pattern recognition, artificial neural networks, etc., in the development of multi-analyte sensing devices. Some of the current research involves new materials for sensors, multi-analyte sensors, process monitoring with sensors, nanosensors, signal processing and enzyme-based sensors.

-R.Ince and R.Narayanaswamy, *Analysis of the performance of interferometry, surface plasmon resonance and luminescence as biosensors and chemosensors*, Analytica Chimica Acta, 2006, In press.

Date submitted: 16th August 2006

Dirk U. Näther, Ph.D.

Edinburgh Instruments Ltd.,
2 Bain Square,
Livingston, EH54 7DQ,
Scotland, UK.
Tel: +44 (0) 150 642 5300   Fax: +44 (0) 150 642 5320
dnather@edinst.com
www.edinburghinstruments.com   &   www.edinst.com

Specialty Keywords: Fluorescence Spectrometers,
Single Photon Counting, Instrumentation.

Dr. Nather has been active in the field of fluorescence since 1989; development engineer in Edinburgh Instruments since 1991. In charge of instrumentation development for the Analytical Instruments Division of Edinburgh Instruments since 2000. Divisional manager of the Analytical Instrument Division since 2006.

Date submitted: Editor Retained.

Miloš Nepraš, Ph.D.

Department of Organic Technology,
University of Pardubice,
Studentská 95, 532 10 Pardubice,
Czech Republic.
Tel: +420 46 603 8500   Fax: +420 46 603 8004
Milos.Nepras@upce.cz

Specialty Keywords: Fluorescent Probes, Bifluorophoric Systems, Structure and Fluorescence Characteristics.

Syntheses and study of relationships between electronic structure and luminescence properties of polynuclear aromatic ketones and quinones and their derivatives. Syntheses and fluorescence characteristics (spectra, quantum yield, fluorescence decay kinetics and solvent effect) of new fluorescent probes derived form acyl and triazinyl derivatives of pyrene, aminopyrenes and aminobenzanthrones. Study of the excitation energy transfer at bifluorophoric systems created form the 3-aminobenzanthrone and aromatic hydrocarbon subsystems.

-V. Fidler, P. Kapusta, M. Nepraš, J. Schroeder, I. V. Rubtsov and K. Yoshihara Femtosecond Fluorescence Anisotropy Kinetics as a Signature of Ultrafast Electronic Energy Transfer in Bichromophoric Molecules Z. Phys. Chem. 216 (2002) 589 – 603.

Date submitted: 29th April 2005

Anne K. Neumeyr-Heidenthal, Ph.D.

Chroma Technology Corp.,
10 Imtec Lane,
Rockingham, VT 05101,
USA.
Tel: 800 824 7662 / +1 802 428 2568   Fax: +1 802 428 2525
anh@chroma.com
www.chroma.com

Specialty Keywords: Fluorescence Microscopy, In-Vivo Imaging, Applications Support.

Dr. Neumeyr-Heidenthal has great comprehensive experience in demonstration, training and support in the fluorescence imaging field, including live-cell/ratio/deconvolution/confocal microscopy and in-vivo imaging.

Currently applications scientist designing and trouble-shooting optics and experimental designs in fluorescence imaging.

Nighswander-Rempel, S. P.
Niles, W. D.

Date submitted: 12th October 2006

Stephen P. Nighswander-Rempel, Ph.D.

Centre for Biophotonics, Dept. of Physics,
University of Queensland,
Brisbane, QLD, 4072,
Australia.
Tel: (61 7) 3365 2816
snighrem@physics.uq.edu.au
www.physics.uq.edu.au/people/snighrem

Specialty Keywords: Melanin, Tissue optics, Imaging.
AIM 2005 = 12.5

My research focuses on optical properties of tissues and biomedical compounds. My Ph.D. research developed a novel spectroscopic imaging technique for diagnosing ischemia in heart tissue, using hemoglobin absorbance. My post-doctoral work studied absorbance and fluorescence of natural and synthetic melanins, using steady-state and time-resolved fluorescence and confocal microscopy.

-*Journal of Molecular and Cellular Cardiology*, 37 (5): 947-957 (2004).

-*Journal of Chemical Physics,* 123 (19): 194901 (2005).

Date submitted: Editor Retained.

Walter D. Niles, Ph.D.

Genoptix Inc., Systems and Applications,
3398 Carmel Mountain Rd., San Diego,
CA, 92037,
USA.
Tel: 858 523 5059   Fax: 858 523 5070
wniles@genoptix.com

Specialty Keywords: Radiometric Imaging, Energy Transfer,
Membrane Dynamics.

Developed quantitative fluorescence resonance energy transfer imaging of membrane dynamics in model and biological systems for understanding essential biophysical mechanisms. Now applying novel fluorescence and optical micromechanics for development of assay technologies (biologies and instrumentation) for drug discovery and diagnostics.

-Endothelial cell-surface gp60 activates vesicle formation and trafficking via Gi-coupled Src kinase signaling pathway. 2000. Journal of Cell Biology 150:1057-1069.
-Radiometric calibration of a video fluorescence microscope for the quantitative imaging of resonance energy transfer. 1995. Review of Scientific Instruments. 66:3527-3536.

Date submitted: Editor Retained.

Christopher G. Norey, Ph.D.

Amersham Biosciences, The Maynard Centre,
Forest Farm, Whitchurch,
Cardiff, CF14 7YT,
Wales, UK.
Tel: +44 (0) 292 052 6439   Fax: +44 (0) 292 052 6230
christopher.norey@amersham.com
www.amershambiosciences.com

Specialty Keywords: Polarization, HTS Instrumentation, Assays.

Our interests are development of systems relevant for high throughput screening assays, employing fluorescence polarization, FRET and time resolved-FRET techniques. Primarily using CyDye™ fluors and Eu (TMT) chelates with detection via single well PMT readers or whole plate imaging platforms, such as LEADseeker™ multi-modality imaging system. We have a particular interest in receptor ligand interactions, protease cleavage and kinase assays. Recently we have been investigating the application of fluorescence lifetime to these areas.

A. Fowler, I. Davies and C. Norey, (2000), A Multi-Modality Assay Platform for Ultra-High Throughput Screening. *Current Pharmaceutical Biotechnology*, 1, 265-281.

A. Harris, S. Cox and C. Norey, (2002), High-throughput fluorescence polarization receptor binding assays. In: *LifeScience News*, Amersham Biosciences UK Limited, issue 10, 17-19.

Date submitted: 1ˢᵗ June 2005

Mercedes Novo, Ph.D.

Universidad de Santiago de Compostela,
Facultad de Ciencias, Departamento de Química Física,
Campus Universitario s/n, E-27002 Lugo,
Spain.
Tel: +34 98 228 5900   Fax: +34 98 228 5872
mnovo@lugo.usc.es

Specialty Keywords: Fluorescence, Data Analysis, FCS.
AIM 2005 = 14.6

Study of the influence of confined media on proton transfer and charge transfer processes. Fluorescent probes for the characterisation of supramolecular structures. Development and implementation of new data analysis methods for steady state and time resolved fluorescence data.

-W. Al-Soufi, B. Reija, M. Novo, S. Felekyan, R. Kühnemuth, and C.A.M. Seidel (2005) Fluorescence Correlation Spectroscopy, a Tool to investigate Supramolecular Dynamics: Inclusion Complexes of Pyronines with Cyclodextrin. *JACS*, 127, 8775-8784.

-B. Reija, W. Al-Soufi, M. Novo, J. Vázquez Tato (2005) Specific interactions in the inclusion complexes of Pyronines Y and B with □-cyclodextrin, *J. Phys. Chem. B*, 109, 1364-1370.

Date submitted: Editor Retained.

Guillermo Orellana, Ph.D.

Laboratory of Applied Photochemistry, Faculty of Chemistry,
Universidad Complutense Madrid,
28040 Madrid,
Spain.
Tel: +34 91 394 4220   Fax: +34 91 394 4103
orellana@quim.ucm.es
www.ucm.es

Specialty Keywords: Indicator Design, Fiber-optic Sensors, Environmental Analysis and Process Control.

Our currents areas of research are (i) design and fabrication of micro-probes based on molecularly engineered luminescent dyes, novel photochemical reactions and *fiber-optic chemosensors* for in situ analysis of environmental, industrial, and medical parameters, and (ii) synthesis and characterization of nano-probes to investigate the structure of nucleic acids and design artificial photonucleases, The realization of both goals rests on *tailored* luminescent transition metal complexes and organic heterocyclic structures.

-F. Navarro-Villoslada, G. Orellana, M.C. Moreno-Bondi, T. Vick, M. Driver, G. Hildebrand and K. Liefeith, *Anal. Chem.* 2001, *73*, 5150-5156.
-M.E. Jiménez, G. Orellana, F. Montero and M.T. Portolés, *Photochem. Photobiol.* 2000, *72*, 28-34.

Date submitted: 5th July 2005

Uwe Ortmann, M.Sc.

PicoQuant GmbH.,
Rudower Chaussee 29,
Berlin 12489,
Germany.
Fax: +49 (0)30 6392 6567
ortmann@pq.fta-berlin.de
www.picoquant.com

Specialty Keywords: Pulsed Lasers, Time-resolved Spectroscopy, Single Molecule Detection.

Current Status: Head of Systems and Sales / Marketing divisions of PicoQuant GmbH.

Major activities are based on the design and further development of fluorescence lifetime systems, especially in the field of time-resolved photon counting equipment and single molecule detection.

-Application of sub-ns pulsed LEDs in fluorescence lifetime spectroscopy, Proceedings of SPIE, Vol.4648, p.171-178 (2002).

Date submitted: 1st August 2004

Martin H. Otz, (Ph.D. Student)

Syracuse University, Dept. of Earth Sciences,
313 Heroy Geology Laboratory, Syracuse,
Onondaga, 13244-1070,
USA.
Tel: 315 857 4614
mhotz@syr.edu
web.syr.edu/~mhotz/index.html

Specialty Keywords: Dye-tracing, Contaminant Hydrology, Fluorescent Dyes.

A major problem in hydrology is to determine the flow paths of water in organic-rich environments. My research focuses on the use of intrinsic fluorescence to locate organic contaminant plumes. Additionally I developed dye-tracing techniques using organic fluorescent dyes to successfully trace water flow paths in heavily contaminated aquifers.

-Otz, M.H., Hinchey, E., and Siegel, D.I., 2003, Using synchronous spectro-fluorometry for tracing oil-contaminated water under an inaccessible factory [abs.]: Transactions Geological Society of America, v. 35, no. 6, p. 413.

-Otz, M.H., Otz, H.K., Ines Otz, and Siegel, D.I., 2003, Surface water/groundwater interaction in the Piora Aquifer, Switzerland: evidence from dye tracing tests: Hydrogeology Journal, v. 11, no. 2, p. 228-239.

Date submitted: 13th July 2004

Tara Chandra Pant, Ph.D.

Photophysics Laboratory, Department of Physics,
Kumaon University, Naninital,
Uttranchal,
India.
PIN 236001 Tel.: +91 594 223 7450   Fax: +91 594 223 5576
tc_pant@yahoo.com

Specialty Keywords: Fluorescence, Excitation Energy Transfer, Optical Sensors.

Electronic Excitation Energy Transfer and Migration in organic dyes and rare earths in solution, glassy and polymeric media. Application of Fluorescence Resonance Energy transfer (FRET) as optical sensors and Luminescence solar collectors. Excitation Energy transfers in micro droplets using techniques of Fluorescence Microscopy.

-An optical approach for sensing pH based on energy transfer in Nafion matrix.V. Misra, H. Mishra, H. C. Joshi and T.C. Pant; Sensor and Actuators B: Chemical, 82 (2002) 133-142.
-Fluorescence studies of Salicylic acid doped in polyvinal alcoho film as a water/ humidity sensor. H. Mishra, V. Misra, M. S. Mehta, T. C. Pant and H. B. Tripathi, J. Physical Chemistry: A 108 (2004) 2346-2352.

Pantano, P.
Papageorgiou, G. C.

Date submitted: 28th June 2005

Paul Pantano, Ph.D.

Department of Chemistry,
The University of Texas at Dallas,
Richardson, TX 75083-0688,
USA.
Tel: 972 883 6226   Fax: 972 883 2925
pantano@utdallas.edu
www.utdallas.edu/dept/chemistry/faculty/pantano.html

PantanoLABO
est. 1996

Specialty Keywords: Microarrays, Sensors, Cell Adhesion.

The driving force of this research laboratory is the development of elegant analytical techniques and methodologies to understand complex chemical systems.   Areas of expertise include chemical microscopy, modified protein surfaces, optical and electrochemical sensors and arrays, and *in situ* biological measurements including a decade of experience designing novel sensors for monitoring the oxidative stress response of single cells and tissue. Current research involves characterizing the biocompatibility and potential cytotoxicity of nanoparticles.

Date submitted: 26th July 2005

George C. Papageorgiou, Ph.D.

National Center of Scientific Researvh Demokritos,
Institute of Biology,
Athens 15310,
Greece.
Tel: +30 210 650 3551   Fax: +30 210 651 167
gcpap@bio.demokritos.gr   gcpap@ath.forthnet.gr

Specialty Keywords: Photosynthesis, Chlorophyll, Fluorescence.

Chlorophyll *a* (Chl *a*) fluorescence informs on photon capture, excitation transfer and trapping, and electron transport, all key processes of oxygenic photosynthesis occuring in the thylakoid membrane (1). In cyanobacteria, however, it informs further on osmotic cell volume changes, because cytoplasmic osmolality regulates how phycobilisome excitation partitions between photosystems II and I. This allows the use of Chl *a* fluorescence to probe cell membrane properties, such as water and solute transport at time resolutions of approx. 1 ms (2).

-G. C. Papageorgiou (2004) Fluorescence of photosynthetic pigments *in vitro* and *in vivo*. In G. C. Papageorgiou and Govindjee (Eds) *Chlorophyll Fluorescence*: *A Signature of Photosynthesis*, pp. 43 - 63. Springer, Dordrecht.

Date submitted: 8th August 2005

Dmitri B. Papkovsky, Ph.D.

Biochemistry Department,
University College Cork,
Lee Maltings, Prospect Row, Cork,
Ireland.
Tel: / Fax: +353 21 490 4257
d.papkovsky@ucc.ie
www.ucc.ie/ucc/depts/biochemistry/staff/dpapkov.html

Specialty Keywords: Phosphorescence, Time-Resolved Fluorescence, Porphyrins, Probes, Oxygen Sensing.
AIM 2005 = 20.2

Research interests are in time-resolved fluorescence, room temperature phosphorescence, quenched-phosphorescence oxygen sensing and respirometry; two-photon excitation fluorescence. Focusing on analytical biochemistry, development of phosphorescent metalloporphyrin labels, probes and bioconjugates; TR-F bioaffinity assay, separation-free and micro-volume assays and screening systems, new biomarkers, fluorescent imaging systems and live cell imaging.

-Papkovsky D.B. - *Screening: Trends in Drug Discovery*, GIT Verlag, 2005, **6**: 46-47.
-O'Mahony F.C. et al. - *Environm. Sci. Technol.*, 2005, 39: 5010-5014.

Date submitted: 5th April 2006

Abraham H. Parola, Ph.D.

The Department of Chemistry,
Dean, Faculty of Natural Sciences
Ben-Gurion University, P.O. Box 653, Beer Sheva,
Israel, 84105.
Tel: Fax Numbers
Tel: 97 28 647 2454   Fax: 97 28 647 2943
aparola@bgu.ac.il

Specialty Keywords: Membrane Dynamics, Protein & Protein-Ligand / Drug Interactions, Time Resolved Fluorescence.

Research topics: The role of hydrophobic interactions in membrane and non-membrane protein function and regulation, signal transduction, cell cycle and proliferation, cell differentiation and intercellular interactions, angiogenesis, apoptosis, magnetic field effects on biological systems.

-Radical scavengers suppress low frequency EMF enhanced proliferation in cultured cells and stress effects in higher plants. A.H. Parola, D. Kost, G. Katsir, E. Ben-Izhak Monselise and R. Cohen-Luria. *The Environmentalist* 25, 103-111 (2005).

-Membrane-catalyzed nucleotide exchange on DnaA: Effect of surface molecular crowding. Aranovich A., G.Y. Gdalevsky, R. Cohen-Luria, I. Fishov, and A.H. Parola. Accepted for publication in the *J. Biol. Chem.* M5:10266 (0306).

Date submitted: 5th April 2006

Digambara Patra, Ph.D.

Department of Physics, Kinosita Lab,
School of Science and Engineering,
Waseda University, Okubo 3-4-1, Shinjuku-ku,
Tokyo 169-8555.
Tel: 0081 35 952 5871   Fax: 0081 35 952 5877
digpatra@rediffmail.com

Specialty Keywords: Fluorescence Spectroscopy &
Microscopy, Single Molecule Detection, Biophysics.
AIM 2005 = 12.7

Optical techniques have made the possibilities of identifying and manipulating chemical/biological phenomenon at single molecular level. Evaluating angular dipole orientation of molecules at single molecular level not only provides the fundamentally very important characteristics of emission properties[1] but probes the individual molecular events in a multi-fluorophoric system[2]. Visualization of rotational motion of a single rotary motor like F1-ATPase during ATP hydrolysis can directly be achieved under a microscope. Coupling this method with magnetic tweezers system would be crucial to explore the possibility of ATP synthesis by F1-ATPase.

-Digambara Patra, Ingo Gregor, Jörg Enderlein, Markus Sauer (2005) 'Defocused imaging of quantum-dot angular distribution of radiation', 87, 101103.

Date submitted: 17th October 2006

Leonid D. Patsenker, Ph.D.

Department of Organic Luminophores and Dyes,
State Scientific Institution "Institute for Single Crystals",
of the National Academy of Sciences of Ukraine,
60, Lenin Ave., Kharkov, 61001, Ukraine.
Tel: +38 057 341 0272   Fax: +38 057 340 9343
patsenker@isc.kharkov.com
www.isc.kharkov.com/old/patsenker

Specialty Keywords: Synthesis, Investigation, Application.
Current Interests: Synthesis, experimental investigation, quantum chemical simulation, and application of organic luminophores and dyes; development of fluorescent labels and probes for biomedical applications, organic scintillators, microspheres and classification dyes for multiplexed analysis systems. Classes of compounds are cyanines, squaraines, diarylazoles, pyrazolines, fused aromatic and heterocyclic systems such as naphthalic and perylenetetracarboxylic acid derivatives, indoles, and quinazoliniun salts among others.

-Ioffe VM. Gorbenko GP, Domanov YA,et al. A new fluorescent squaraine probe for the measurement of membrane polarity Journal of Fluorescence 16 (1): 47-52 Jan 2006
-Tatarets AL, Fedyunyayeva IA Dyubko TS, et al. Synthesis of water-soluble, ring-substituted squaraine dyes & their evaluation as fluorescent probes and labels. Analytica Chimica Acta. 570(2):214-223 June 16 2006.

Date submitted: 5th April 2006

Albertha (Bert) Paul, M.S.

Boston Electronics Corporation,
91 Boylston Street, Brookline,
MA 02445,
USA.
Tel: 800 347 5445 / 617 566 3821   Fax: 617 731 0935
bpaul@boselec.com
www.boselec.com

Specialty Keywords: TCSPC, Spectroscopy, Photochemistry.

Applications Engineer at Boston Electronics Corporation, North American agents for Edinburgh Instruments Ltd of Edinburgh, Scotland and for Becker & Hickl GmbH of Berlin, Germany. Specialist in photochemistry.

Date submitted: 5th April 2006

William H. Pearson, Ph.D.

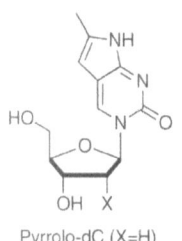

BERRY&ASSOCIATES

Pyrrolo-dC (X=H)
Pyrrolo-C (X=OH)

Berry & Associates, Inc.,
2434 Bishop Circle East,
Dexter, MI 48130,
USA.
Tel: 800 357 1145   Fax: 734 426 9077
wpearson@berryassoc.com
www.berryassoc.com

Specialty Keywords: Fluorophores, Dark Quenchers, Fluorescent Nucleosides.

We are a leading source of nucleosides, modified nucleosides, DNA/RNA synthesis reagents (phosphoramidites, CPGs), as well as fluorophores and quenchers. We offer fluorescence quenchers, carboxyfluoresceins, carboxytetramethylrhodamines, and fluorescent nucleosides, all in forms suitable for incorporation into nucleic acids or peptides. Our current efforts include the development of new fluorophores and dark quenchers. We would also be interested in discussing your custom fluorescence needs.

-"Pyrrolo-dC and Pyrrolo-C: Fluorescent analogs of cytidine and 2'-deoxycytidine for the study of oligonucleotides," D. A. Berry, et. al, *Tetrahedron Letters*, 2004, *45*, 2457-2461.

Peknicova, J.
Peltié, P.

Date submitted: 15th July 2005

Jana Peknicova, Ph.D.

Department of Biology and Biochemistry of Fertilization,
Videnska 1083, Prague 4,
142 20,
Czech Republic.
Fax: +420 244 471 707
jpeknic@biomed.cas.cz
www.img.cas.cz/dbbf

Specialty Keywords: Biology of Reproduction, Fertilization, Human Infertility.

Selected endocrine disruptors (present as a food contaminants with constitute reproductive toxicology risk) were tested in the influence on in *vivo* fertility of outbread lines of mice. Diethylstilbestrol had a negative influence on fertilization, contrary to that, treatment by phytoestrogens (genistein) had not affected on reproduction. Monoclonal antibodies against intra-acrosomal sperm proteins were used as suitable and sensitive markers for detection of the state of acrosome, sperm quality, sperm pathology and spermatogenesis in the men with azoospermia.

-Peknicova J., Chladek D., Hozak P.: Monoclonal antibodies against sperm intra-acrosomal antigens as markers for male infertility diagnostics and estimation of spermatogenesis. Amer. J. Reprod. Immunol.., 53:42-49, 2005.

Date submitted: 4th August 2004

Philippe Peltié, Ph.D.

CEA/Grenoble,
Dept.Techno.Biologie Santé, 17,rue des Martyrs,
38054 Grenoble cedex9,
France.
Tel : 33 43 878 2415 Fax : 33 43 878 5787
philippe.peltie@cea.fr

Specialty Keywords: Fluorescence Instrumentation, In Vivo Fluorescence Imaging.

We have developed during 6 years fluorescent DNA chips readers; my work, now, focuses on the development of instrumentation for in vivo fluorescence imaging. Fluorescent probes can be functionnalized to target specific organs, lesions, tumor. This non-invasive technique makes it possible to localize and measure tumors in small animals; this technique may be extended to human diagnosis for shallow organs or lesions.

-Fluorescence detection for DNA chips and labs on chips and perspective for integrated systems. IXth international symposium on luminescence spectrometry in biomedical and environmental Analysis; may 15-17, 2002; Montpellier, France.

Date submitted: Editor Retained.

Michael J. Pender, M.S.

Nanochron LLC.,
4201 Wilson Blvd. #110-615,
Arlington, Virginia 22203, USA.

Michael.Pender@Nanochron.com
www.nanochron.com

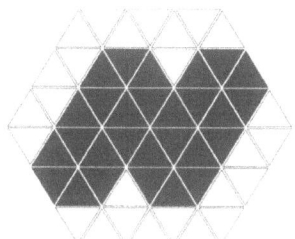

Specialty Keywords: Photonics, Predictive Modeling.

My work focuses on the development of application-specific optical devices. Specific topics include intra-molecular photonic transfer in fluorescent and quasi-fluorescent optical channels and predictive modeling of the properties of fluorophores in photonic devices for optical communications and signal processing.

-M. Pender (2001). Optical matrix photonic logic device and method for producing the same, Patent Cooperation Treaty Application No. PCT/IB01/00888.

Date submitted: 4th May 2006

Xinzhan Peng, Ph.D.

LI-COR Biosciences,
4647 Superior Street,
Lincoln, Nebraska 68504,
USA.
Tel: 402 467 0796   Fax: 402 467 0819
Xinzhan.peng@licor.com
www.licor.com

Specialty Keywords: Near-infrared fluorescence and quencher dyes; Protease assays; Fluorescence resonance energy transfer.
AIM 2005 = 6.5

My current research focuses on the design and development of novel near-infrared (NIR) fluorescent dyes and probes; design and synthesis of NIR quencher dyes; development of NIR-FRET based protease assay platform for drug screening targeting HIV-1 protease, bata-secretase, etc.

-Xinzhan Peng, Daniel. R. Draney, Jiyan Chen, William M. Volcheck (2006). "Phthalocyanine dyes". US 7,005,518 B2.
-Xinzhan Peng*, Daniel R. Draney, William M. Volcheck (2006). "Quenched near-infrared fluorescent peptide substrate for HIV-1 protease assay". *Proceedings of SPIE* Vol. 6097, 60970F.

Penzkofer, A.
Perez-Inestrosa, E.

Date submitted: 18th April 2006

Alfons Penzkofer, Ph.D.

Naturwissenschaftliche Fakultät II – Physik,
Universität Regensburg,
Universitätsstraße 31, Regensburg, D-93053,
Germany.
Tel: +49 0941 943 2107   Fax: +49 0941 943 2754
alfons.penzkofer@physik.uni-regensburg.de
www.physik.uni-regensburg.de/forschung/penzkofer

Specialty Keywords: Blue-Light Receptors, Luminescent Polymers, Ultrafast Spectroscopy.

We undertake an absorption and spectroscopic characterisation of organic molecules, sensory biological photo-receptors (flavin chromophores) and luminescent polymers by linear, nonlinear, and time resolved optical methods. We perform laser studies on thin-film luminescent polymers and solid-state dye lasers.

-P. Zirak, A. Penzkofer, T. Schiereis, P. Hegemann, A. Jung, I. Schlichting, Photodynamics of the small BLUF protein BlrB from Rhodobacter sphaeroides, J. Photochem. Photobiol. B: Biology, 180 (2006) 180-196.

-W. Holzer, A. Penzkofer, T. Tsuboi, Absorption and Emission Spectroscopic Characterization of Ir(ppy)3, Chem. Phys. 308 (2005) 93-102.

Date submitted: 8th June 2005

Ezequiel Perez-Inestrosa, Ph.D.

Organic Chemistry,
University of Malaga,
Campus Teatinos, Malaga, 29071,
Spain.
Tel: 34 95 213 7565   Fax: 34 95 213 1941
inestrosa@uma.es

Specialty Keywords: Acceptor-Donor Molecular Devices,
Molecular Recognition.
AIM 2004 = 12.5

Fluorescence has been applied to the study of the metal cation binding of photoresponsive complexing systems constituted of bisarylcyclophanes, able to form 1:1 and 1:2 complexes. The metal cation binding ability in the S1 is shown to diminish and it is interpreted as a transitory photodecomplexation between the metal cations and the phenolic oxygen atoms. Determination of the association constants allow a discussion of the cooperative effects found in the ground and excited state.

-J.-M. Montenegro, E. Perez-Inestrosa, D. Collado, Y. Vida and R. Suau (2004). A natural-product-inspired photonic logia gate based on photoinduced electron-transfer-generated dual-channel fluorescence *Organic Letters* 6(14), 2353-2355.

Date submitted: 11th August 2005

Ammasi Periasamy, Ph.D.

Keck Center for Cellular Imaging,
University of Virginia,
Biology, Gilmer Hall 064,
Charlottesville, VA 22904, USA.
Tel: 434 243 7602   Fax: 434 982 5210
ap3t@virginia.edu
www.kcci.virginia.edu/Contact/peri.php

Specialty Keywords: FRET, FLIM, Fluorescence Microscopy.
AIM 2003 = 44.4

Dr. Periasamy created an imaging center at the University of Varginia and developed (orintegrated) number of state-of –the-art light microscopy imaging systems including FLIM, FRET, TIRF and multiphoton microscopy systems. A key area of his research is focused on to study / monitor various biological and clinical systems ranging from a single cell to single molecule in living cells and tissues. Edited two book's, chair person for the multiphoton microscopy conference and running an annual workshop on FRET microscopy.

-H. Wallrabe, and A. Periasamy (2005). FRET-FLIM microscopy and spectroscopy in the biomedical sciences. Curr. Opin. Biotech. 16, 19-27.

-Y. Chen, M. Elangovan, and A. Periasamy (2005) in A. Periasamy and R. N. Day (Eds.), *Molecular Imaging: FRET Microscopy and Spectroscopy*, Oxford University Press, New York, pp. 126-145.

Date submitted: 29th April 2005

Frederick S. Perry.

Boston Electronics Corporation,
91 Boylston Street, Brookline,
MA 02445,
USA.
Tel: 800 347 5445 / 617 566 3821   Fax: 617 731 0935
fsp@boselec.com
www.boselec.com

Specialty Keywords: TCSPC, Photodetection, Spectroscopy.

Frederick S. Perry is the President and founder of Boston Electronics Corporation, North American agents for Becker & Hickl GmbH of Berlin, Germany and for Edinburgh Instruments Ltd of Edinburgh, Scotland.

Specialist in photodetection and in signal processing electronics for photodetection.

Petersen, N. O.
Pispisa, B.

Date submitted: Editor Retained.

Nils O. Petersen, Ph.D.

Department of Chemistry, The University of Western Ontario,
1151 Richmond Street N.,
London, Ontario, N6A 5B7,
Canada.
Tel: 519 661 3138   Fax: 519 661 3139
petersen@uwo.ca
www.uwo.ca/chem

Specialty Keywords: Microscopy, Confocal, Correlation spectroscopy.

Image Correlation Spectroscopy: measurements of density of molecular clusters or single molecules, the degree of aggregation and the extent of association of different molecules into co-localized domains.   Fluorescence photobleaching or fluorescence correlation spectroscopy. Fluorescence measurements in small volumes. Multiphoton excitation in confocal microscopy applications. Protein-protein interactions in domains in membranes of cells. Atomic force microscopy and time-of-flight secondary ion mass spectrometry of membranes and monolayers.
-N.O. Petersen AFCS and Spatial Correlations on Biological Surfaces@ Ch. 8 in AFluorescence Correlation Spectroscopy@ Edited by R. Rigler and E.L. Elson, Springer Verlag (2000).
-C.L. Lee and N.O. Petersen "The Lateral Diffusion of Selectively Aggregated Peptides in Giant Unilamellar Vesicles" Biophysical J. 84, 1756-64 (2003).

Date Submitted: Editor Retained.

Basilio Pispisa, Ph.D.

Dept.of Chemical Sciences and Technologies,
University of Roma Tor Vergata,
Via Ricerca Scientifica, Rome,
00133 Italy.
Tel: +39 067 259 4467   Fax: +39 06 202 0420
pispisa@stc.uniroma2.it
www.stc.uniroma2.it/files/Pispisa%20files/B.Pispisa

Speciality keywords: Biophysical Chemistry, Spectroscopy, Conformational Analysis.

Three major research topics are pursued in the Professor Pispisa's laboratory:
- Structure and molecular dynamics of oligopeptides and polypeptides in solution, mimicking proteins and bioactive compounds;
- Structure-reactivity relationships in model compounds of enzymic materials;
- Structural features of glycopeptides and functionalized peptides in solution and in membranes.

-B. Pispisa et al. (2000) *Biopolymers*, 54, 127-136. 2002 Peptide-Sandwiched Protoporphyrin Compounds Mimicking Hemoprotein Structures in Solution.

-B. Pispisa et al. (2002) *J. Phys. Chem.* B 106, 5733-5738. Effects of Helical Distortions on the optical Properties of Amide NH Infrared Absorption in Short Peptides in Solution.

Date submitted: 16th April 2006

Vasyl G. Pivovarenko, Ph.D.

Chemistry Department,
National Taras Shevchenko University of Kyiv,
Volodymyrska 64,
Kyiv 01033, Ukraine.
Tel: +38 044 239 3312   Fax: +380 44 220 8391
pvg@univ.kiev.ua   &   pvg_org@mail.ru
home.chem.univ.kiev.ua/~pvg/
Specialty Keywords: Organic Synthesis, Fluorescence
Spectroscopy, Fluorescence Probes Design.
AIM 2005 = 16.6

Starting from 1994 a series of fluorescence ratiometric sensors based on 3-hydroxyflavones, 3-hydroxyquinolones, crown-ketocyanines, diflavonols and dicyclopentano[b,e]pyridines were synthesized and studied jointly with several groups of scientists. The probes developed were shown to have record sensitivity by their fluorescence parameters on the polarity, hydration or pH of a medium, or on the size of metal cation, or on the state and composition of lipid membrane. The main interests at present time are the probes for cell membrane study, wide-range pH indicators and adenosine triphosphate sensors.

-Shynkar V.V., Klymchenko A.S., Mely Y., Duportail G., Pivovarenko V.G.. Anion Formation of Flavonol Probe in Phosphatidylglycerol Vesicles Induced by HEPES Buffer: A Steady-State and Time-Resolved Fluorescence Investigation. *J. Phys. Chem. B* (2004), *108,* 18750-18755.

Date submitted: Editor Retained.

Emmanuelle Plantin-Carrenard, Ph.D.

Laboratoire de Biochimie Générale et de Glycobiologie,
UFR des Sciences Pharmaceutiques et Biologiques,
Uni. René Descartes - Paris 5, 4 avenue de L'Observatoire,
75006 Paris - France.
Tel: 33 15 373 9652   Fax: 33 15 373 9655
eplantin@wanadoo.fr

Specialty Keywords: Fluorescence Probes, Oxidative Stress,
Apoptose.

Oxidative stress is defined as the pathological outcome of overproduction of oxidative species that overwhelms the cellular antioxidant capacity. The consequence of induced-oxidative stress are studied *in vitro* on adherent and non-adherent cell models. Fluorescent probes are interesting tools to measure with high sensitivity and specificity the modifications of cellular fonctions under oxidant conditions : reactive oxygen species production, modulation of intracellular thiol levels, necrosis/apoptosis balance, cellular adhesion, evaluation of the protective effects of some antioxidant compounds.

-Plantin-Carrenard E. et al. Journal of Fluorescence, 2000; 10 : 167-73.
-Plantin-Carrenard E. et al. Cell Biol Toxicol, 2003; 19 : 121-33.

Date submitted: Editor Retained.

Jaromír Plášek, Ph.D.

Faculty of Mathematics and Physics – Institute of Physics,
Charles University, Ke Karlovu 5,
Prague, CZ-12116,
Czech Republic.
Tel: +420 22 191 1349
plasek@karlov.mff.cuni.cz

Specialty Keywords: Membrane Potential, Polarized Fluorescence, Microfluorimetry.

Research Interests: Lipid order in cell membranes from polarized fluorescence of membrane probes. Fluorescent probing of cell membrane and mitochondrial membrane potential in living cells. ATP binging to a N-domain in the cytoplasmic loop of Na,K-ATPase from binding assays with TNP-ATP.

-J. Plášek and K. Sigler (1996) Slow fluorescent indicators of membrane potential: a survey of different approaches to probe response analysis. *J. Photochem. Photobiol. B: Biology* 33, 101-124.

-D. Gášková, R. Čadek, R. Chaloupka, J. Plášek and K. Sigler (2001) Factors underlying membrane potential-dependent and -independent fluorescence responses of potentiometric dyes in stressed cells: diS-C$_3$(3) in yeast. *Biochim. Biophys. Acta* 1511, 74-79.

Date submitted: 21st May 2006

Manuel Prieto, Ph.D.

Centro de Química-Física Molecular, IST,
Av. Rovisco Pais,
1049-001 Lisbon,
Portugal.
Tel: +35 121 841 9219   Fax: +35 121 846 4457
prieto@alfa.ist.utl.pt

Specialty Keywords: FRET, lipid domains, lipid protein-interaction, lipid-DNA complexes.
AIM 2004 = 13.8

Current Research Interests: Application of steady-state and time-resolved photophysical methodologies to the detection, characterization and dynamics of membrane heterogeneities (domains/rafts). Topology and dynamics of protein/peptide and polyene antibiotics interaction with model systems of membranes. Cholesterol organization in membranes. Lipid-DNA complexes.

-Absence of clustering of phosphatidylinositol-(4,5)-bisphosphate in fluid phosphatidylcholine. J. Lipid Res. 2006.

-Ceramide-platform formation and induced biophysical changes in a fluid phospholipid membrane. *Mol. Membr. Biol.* 23(2), 137 – 148; Competitive binding of cholesterol and ergosterol to the polyene antibiotic nystatin.A fluorescence study. *Biophys. J.* 90 (10), 3625-3631

Date submitted: 16th July 2004

Karel Procházka, Ph.D., D.Sc.

Dept. of Physical and Macromolecular Chemistry,
Faculty of Science, Charles University in Prague,
Albertov 6, 128 43 Prague 2, Prague,
Czech Republic.
Tel.: 001 420 22 195 1293   Fax: 001 420 22 491 9752
prochaz@vivien.natur.cuni.cz
natur.cuni.cz/pmc/group.php?id=5

Specialty Keywords: Time-resolved Fluorescence, Anisotropy, FCS, AFM.

In my group, we study labeled self-assembling polymers by steady-state and time-resolved fluorescence (including anisotropy and FCS) in combination with static and dynamic light scattering, chromatography, electrophoresis, AFM, etc. Results are interpreted with help of original Monte Carlo simulations and self-consistent field calculations. Papers are published in journals oriented on physical chemistry of polymers and colloids (e.g., Macromolecules, Langmuir, J. Phys. Chem., J. Chem. Phys.).

-Matějíček P., Humpolíčková J., Procházka K., et al.: J. Phys. Chem. B 2003, 107, 8232-8240. Matějíček P., Uhlík F., Limpouchová Z., et al (Procházka K.): Macromolecules 2002, 35, 9487-9496.

Date submitted: 8th September 2006

Luca Prodi, Ph.D.

"G. Ciamician" Department,
University of Bologna,
Via Selmi 2, Bologna,
40126, Italy.
Tel: +39 051 209 9481   Fax: +39 051 209 9456
luca.prodi@unibo.it
www.ciam.unibo.it/photochem/prodi.html
Specialty Keywords: Sensors, Nanoparticles, Labels.
AIM 2004 = 30.6

The research activity of Dr. Luca Prodi is documented by more than one hundred papers published on international journals and is located in the field of supramolecular photochemistry. Recently the research has been focussed on the synthesis & characterisation of luminescent labels and sensors. In this context, he also worked on the synthesis and characterisation of metal & silica nanoparticles, in order to obtain more efficient sensors through signal amplification. He is also a co-founder of Cyanagen, a spin-off company, active in the synthesis of new dyes.

Prodi L (2005) Luminescent chemosensors: from molecules to nanoparticles. New J Chem 29:20
-Farruggia G, Iotti S, Prodi L, Montalti M, Zaccheroni N, Savage P B, Trapani V, Sale P, Wolf F I (2006) 8-Hydroxyquinoline Derivatives as Fluorescent Sensors for Magnesium in Living Cells. J Am Chem Soc 128:344.

Date submitted: 31st July 2004

René J. Püschl, M.Sc.

Chemistry Department,
University of Siegen,
Adolf-Reichwein Str. 2 - 4,
D-57068 Siegen, Germany.
Tel: +49 (0) 271 740 4022
pueschl@chemie.uni-siegen.de
www.uni-siegen.de/~ag-drex/

Specialty Keywords: Fluorescent Dyes, Absorption,
Fluorescence, Capillary Electrophoresis, Lab-on-Chip.

My research interest is focused on determining the optical properties of dye-solutions with absorption and fluorescence spectroscopy, e. g. highly precise determination of fluorescence quantum yields, the observation and investigation of dye-aggregation in water, as well as measurements in very dilute solutions.
My second field of interest lies in the area of capillary electrophoresis using fluorescence-based detection techniques (Lab-on-Chip).

Date submitted: 24th October 2006

Kallikat N. Rajasekharan, Ph.D.

Department of Chemistry,
University of Kerala,
Trivandrum, Kerala,
695581 INDIA.
Tel: 91 471 241 8782
kn.rajasekharan@gmail.com
www.keralauniversity.edu./knr/index.html

Specialty Keywords: Fluroprobes, Bioconjugates.

We have developed thiol-directed fluorolabels for protein thiol group that are useful as fluoroprobes and in protein detection on SDS-PAGE. Fluorescent hetaryl amine based enzyme substrate mimics have also been synthesized.

-K. V. D. Babu and K. N. Rajasekharan (2002). Synthesis of 4-(diarylpyrazolinyl)-phenylmaleimides as thiol-directed fluoroprobes. *Synth. Commun.* 32, 1535-1542.

-J. S. Nair and K. N. Rajasekharan (2004). Synthesis and fluorescence properties of 3-benzoxa- and thiazol-2-ylquinoline-5 or 7-maleimides. *Indian J. Chem.* 43B, 1944-5.

Date submitted: 1st September 2006

Krishanu Ray, Ph.D.

Center for Fluorescence Spectroscopy,
Department of Biochemistry & Molecular Biology,
University of Maryland,
725 West Lombard Street, Baltimore 21201, USA.
Tel: 410 706 7500   Fax: 410 706 8408
krishanu@cfs.umbi.umd.edu
cfs.umbi.umd.edu

Specialty Keywords: Single molecule imaging spectroscopy, Fluorescence, Organized ultrathin films.

Single fluorescent molecules as nanoscale optical probes. My current area of research focuses on the application of single molecule spectroscopy in biomedical research and development of a fundamental understanding of metal-enhanced fluorescence and surface plasmon coupled emission.

-K. Ray, R. Badugu, and J. Lakowicz (2006). Metal-enhanced fluorescence from CdTe nanocrystals: A single-molecule fluorescence study. J. Am. Chem. Soc. 128, 8998.

-K. Ray, R. Badugu, and J. Lakowicz (2006). Langmuir-Blodgett monolayers of long-chain NBD derivatives on silver island films: well organized probe layer for the metal enhanced fluorescence studies. J. Phys. Chem. B 110, 13499.

Date submitted: 30th June 2005

M. Elisabete C. D. Real Oliveira, Ph.D.

Physics Department, University of Minho,
Campus de Gualtar, Braga,
Portugal, 4710-057,
Portugal.
Tel: +35 125 360 4325   Fax: +35 125 367 8981
beta@fisica.uminho.pt

Specialty Keywords: Microemulsions, Cationic Vesicles/ Cholesterol, Phospholipids, Silicone-hydrogel Contact Lenses.

In the last year my research has been focused on the to characterization microemulsions using nonionic surfactants the interaction of cationic vesicles (DODAB and DDAB) with neutral phospholipids and cholesterol by fluorescence anisotropy using the fluorescence probes, pyrene , nile red, etc. At the moment I am interested on studying changes in properties of the silicone-hydrogel contact lenses properties after wear (adhesion or proteins, lipids and michoorganisms).

-Transitions in ternary surfactant/alkane/water microemulsions as viewed by fluorescence, *G. Hungerford, M.E.C.D. Real Oliveira, E.M.S. Castanheira, H.D.Burrows and M.da G. Miguel*, Progress in Colloid and Polymer Science,123, 1-4 (2005).

Date submitted: 9th June 2005

Glen I. Redford, Ph.D.

Physics,
University of Illinois, Urbana-Champaign,
1110 W Green, Urbana,
Champaign, 61802, USA.
Tel: 217 333 3054
redford@uiuc.edu
netfiles.uiuc.edu/redford/www/

Specialty Keywords: FLI, Fluorescence, Lifetime.

Dr. Redford's research interests include: Using fast fluorescence lifetime imaging in a variety of applications. Included are applications in cancer detection, plant photosynthesis, and FRET. Glen specializes in the hardware, software, and analysis techniques required to do real-time lifetime imaging.

Date submitted: 8th May 2006

Renata R. Reisfeld, Ph.D.

Inorganic Chemistry, Hebrew University,
Givat Ram, Jerusalem,
91904 Jeruslem,
Israel.
Tel: 97 22 658 5323   Fax: 97 22 658 5319
renata@vms.huji.ac.il
chem.ch.huji.ac.il/~rena/

Specialty Keywords: Fluorescence, Lasers, Nanoparticles, Sol-Gel.

Prof. Reisfeld, DHC: (1) Lyon (France) –1993; (2) Bucharest (Romania) 01998; Wroclaw (Poland) – 2005.

Dr. Reisfeld Current interests also include: Tunable lasers in visible, fluorescence of molecules and lantanides, nanoparticles, sol gel glasses, fluorescent sensors, luminescent complexes of lanthanides, time resolved luminescene of minerals, active waveguides and low reflectance and low dielectric glass films deposited on electronic materials.

-R. Reisfeld, Fluorescent dyes in sol-gel glasses, *J. of Fluorescence,* 12 (314) (2002) 317-325.
Renata Reisfeld, Sol gel processed lasers, *Sol-Gel Technology (Handbook)* Vol. 3, Chap.12, (2004) 239-261, ed. Summio Sakka.

Date submitted: Editor Retained.

Ute Resch-Genger, Ph.D.

Project Group I.3902,
Bundesanstalt für Materialforschung und –prüfung (BAM),
(Federal Institute for Materials Research and Testing),
Richard-Willstätter-Str. 11, D-12489 Berlin,
Germany.
Tel: +49 308 104 1134   Fax: +49 308 104 5005
ute.resch@bam.de
www.bam.de
Specialty Keywords: Fluorescent Standards, Fluorescent Probes
and Sensors, Time Resolved Fluorometry, Quality Assurance.

Current Research Interests: Design and spectroscopic study of functional dyes and fluorescent sensor molecules. Quality assurance and standardization including development of fluorescent standards for steady state and time resolved fluorometry.

-K. Rurack, U. Resch-Genger (2002). Rigidization, preorientation and electronic decoupling – the magic triangle for the design of highly efficient sensors and switches, Chem. Soc. Rev. 31, 116-127.

Date submitted: 13th June 2006

Wolfgang Rettig, Ph.D.

Institut für Chemie der Humboldt-Universität zu Berlin,
Brook-Taylor-str. 2,
12489 Berlin,
Germany.
Tel: +49 30 2093 5585 (Secret. 5561)   Fax : +49 30 2093 5574
rettig@chemie.hu-berlin.de
www.chemie.hu-berlin.de/wr/index.html
Specialty Keywords: Time-resolved fluorescence, Adiabatic
photoreactions, TICT.
AIM 2005 = 23.9

Mechanisms of photochemical primary processes (electron and proton transfer; trans-cis and valence isomerizations; visual process); ultrafast fluorescence and absorption spectroscopy; solvation of excited states; quantum-chemical modelling of photoreactions; fluorescence probes for biology, medicine and analytical chemistry; fluorescence polymer probing. Many studies enriching the field of compounds with anomalous fluorescence properties linked with intramolecular twisting (TICT).

-Hani El-Gezawy, Wolfgang Rettig*, Andrzej Danel Gediminas Jonusauskas, Probing the Photochemical Mechanism in Photoactive Yellow Protein, J. Phys. Chem. B 109 (2005) 18699-18705

Date submitted: 30th June 2005

Alexis Rochas, Ph.D.

idQuantique SA,
Chemin de la Marbrerie 3,
1227 Carouge-Geneva,
Switzerland.
Tel: +41 22 301 8376   Fax: +41 22 301 8379
alexis.rochas@idquantique.com
www.idquantique.com

Specialty Keywords: Single Photon Avalanche Diodes, Fluorescence, Time-correlated single photon counting, imaging.
Research includes the design, the fabrication and the test of single photon detectors for applications in biology and medicine. The detectors and detector arrays are based on single photon avalanche diodes fabricated in industrial CMOS technologies. They exhibits a state-of-the-art timing resolution of less than 40ps.
-M. Gösch, A. Serov, A. Rochas, H. Blom, T. Anhut, P.A. Besse, R.S. Popovic, T. Lasser, and R. Rigler, Parallel Single Molecule Detection with a fully integrated Single Photon 2×2 CMOS-Detector Array, Journal of Biomedical Optics 9(5), 913-921.
-A. Rochas, M. Gösch, A. Serov, P.-A. Besse, and R.S. Popovic, First Fully Integrated 2-D Array of Single-Photon Detectors in Standard CMOS Technology, IEEE Photonics Technology Letters 15(7), 963-965.

Date submitted: 19th May 2006

David E. Roll, Ph.D.

Chemistry Department, Roberts Wesleyan College,
2301 Westside Drive, Rochester,
Monroe County, 14624,
USA.
Tel: 585 594 6485   Fax: 585 594 6482
rolld@roberts.edu

Specialty Keywords: Chlamydia Infection, Topoisomerase, Gold Nanoparticles.
Type I topoisomerase is an enzyme that plays a role in the regulation of DNA supercoiling in the cell. Research in this lab is directed at understanding the role of this enzyme in the initiation of Chlamydia infection in eukaryotic cells and the role that phosphorylation may play in regulating this enzyme's activity. In addition, gold nanoparticles and metal enhanced fluorescence may provide valuable tools for the rapid detection of Chlamydia infections.
-D. Roll, J. Malicka, I. Gryczynski, Z. Gryczynski and J. R. Lakowicz (2003). Metallic colloid wavelength-ratiometric scattering sensors, *Anal. Chem.*, 75(14), 3440-3445.
-J.Zhang, , D. Roll, C. D. Geddes, and J. R. Lakowicz, (2004). Aggregation of silver nanoparticle-dextran adducts with concanavalin A and competitive displacement with glucose, *Journal of Physical Chemistry B*, 108(32), 12210-12214.

Date submitted: Editor Retained.

Ella A. Romodanova, Ph.D.

Department of Biological and Medical Physics,
School of Radiophysics,
V.N.Karazin Kharkov National University,
4 Svobody Sqv., Kharkov, 61077, Ukraine.
Tel: + 380 57 245 7576   Fax: + 380 57 235 3977
dyubko@uaic.net
www-biomedphys.univer.kharkov.ua/

Specialty Keywords: Fluorescence Spectroscopy, Fluorescent Probes, Proteins, Cell Suspensions.

Main research interests: My research interests include fluorescence analysis application to investigation of physical factors (low temperatures, laser and ionizing radiation etc.) action on the biopolymers solutions and cell suspensions. Author and co-author of more then 100 scientific and methodical works.

-Romodanova E.A., Gavrik V.V., Roshal A.D. et al. Changes in HSA Conformation under the Action of Freezing and Laser Radiation as Judged by Fluorescence of Nafthalic Acid Derivative, *Problems of Cryobiology (2002)*, **3**, 28-32.
-Romodanova E.A., Dyubko T.S. et al. MNBIS as marker of protein macrostructure changes, *Biophysical Bulletin (Visn. Khar. Univ.)*, (2002), Ser. Radiophysics and Electronics, Issue 2 (570), 302-307.

Date submitted: 12th September 2006

Alexander D. Roshal, Ph.D.

Institute of Chemistry,
at N.V.Karazin Kharkov National University,
4. Svoboda Square, Kharkov,
61077, Ukraine.
Tel: 38 057 707 5335   Fax: 38 057 707 5130
Alexandre.D.Rochal@univer.kharkov.ua
home.univer.kharkov.ua/rochal

Specialty Keywords: Flavonoids, Flavonoid Complexes, Absorption and Fluorescence Spectroscopy.

Research interests: Structure and physico-chemical properties of flavonoids, Proton transfer in flavonoids under excitation, Complexation of flavonoids in the ground and excited states, Spectral properties and analysis of flavonol complexes, Using natural and synthetic flavonoids, coumarins and relative substances as the fluorescent probes for biochemistry and biophysics.

-A.D. Roshal, M.I. Lvovska, V.P. Khylia. *Russian Journal of Physical Chemistry*, 79 (2005), 1287-1291.
-M.I. Lvovska, A.D. Roshal, A.O. Doroshenko, A.V. Kyrychenko, V.P. Khilya. *Spectrochimica Acta A*, 65 (2006), 397-405

Date submitted: 24th July 2003

Victoria V. Roshchina, Ph.D., D.Sc.

Laboratory of Microspectral Analysis of Cells and Cellular Systems,
Russian Academy of Sciences Institute of Cell Biophysics,
Institutskaya, 3, Pushchino, Moscow Region, 142290, Russia.
Tel: (095) 923 7467, add.293   Fax: 7 (0967) 79 0509
lyudam@icb.psn.ru

Specialty Keywords: Plant Physiology and Biochemistry, Sensory Systems, Spectral Analysis.

Autofluorescence of intact plant microspores, which serve for the vegetative or sexual breeding, has been studied. The emission is changed at the microspores germination. Reactive oxygen species (ozone, superoxide anionradical and peroxides) contribute in the autofluorescence and chemiluminescence of pollen and vegetative microspores.

-V.V.Roshchina, E.V.Melnikova, V.A.Yashin and V.N. Karnaukhov (2002) Autofluorescence of intact spores of horsetail Equisetum arvense L. during their development. Biophysics (Russia)47(2), 318-324.

-V.V.Roshchina, A.V. Miller, V.G. Safronova, and V.N. Karnaukhov (2003) Reactive oxygen species and luminescence of intact cells of microspores. Biophysics (Russia) 48 (2), 259-264.

Date submitted: Editor Retained.

Anatoly N. Rubinov, Ph.D., D.Sc.

Laboratory of Dye Lasers,
Stepanov Institute of Physics,
70 Skorina Avenue, 220072 Minsk, Belarus.
Tel: 375 (17) 284 5624   Fax: 375 (17) 284 1646
rubinov@ifanbel.bas-net.by

Specialty Keywords: Dye Lasers, Ultrafast Spectroscopy, Fluorescent Probes.

The main results are in the field of ruby and neodymium glass lasers (early 1960s); various types of dye lasers and laser dyes (State Prize of USSR, 1972); laser spectroscopy of organic solutions (State Prize of Belarus, 1994); intracavity laser spectroscopy; distributed-feedback (DFB) lasers including holographic DFB lasers; mode locked dye lasers and time resolved laser spectroscopy of organic molecules in solutions and bio-membranes; interaction of gradient laser fields with biological objects.

Date submitted: 20th July 2005

Angelika Rueck, Ph.D.

Institut for Lasertechnologies (ILM),
Helmholtzstrasse 12, Ulm,
89081,
Germany.
Tel: +49 73 114 2916   Fax +49 73 114 2942
angelika.rueck@ilm.uni-ulm.de
www.uni-ulm.de/ilm

Specialty Keywords: FLIM, SLIM, PDT.
AIM 2003 = 3.8

Development of methods for spectral fluorescence lifetime imaging (SLIM), based on time correlated single photon counting in combination with laser scanning microscopes for detection and dynamic analysis of signal transduction pathways in living cells during photodynamic therapy (PDT). Cellular characterization and evaluation of new photosensitizers and photosensitizer metabolites. Definition of protein standards for FLIM/FRET measurements of protein interactions in living cells. Imaging of living cells by confocal raman microscopy.

-M. Kress and A. Rück, Time-resolved microspectrofluorometry and FLIM of photosensitizers using ps pulsed diode lasers in laser scanning microscopes. J. Biomed. Optics,2003, 8(1): 26-32.
-A. Rück et al., PDT with TOOKAD studied in the chorioallantoic membrane of fertilized eggs. Photodiagnosis and Photodynamic Therapy, 2005, 2: 79-90.

Date submitted: 12th October 2006

Knut Rurack, Ph.D.

Div. I.5 "Bioanalysis",
Bundesanstalt für Materialforschung und -prüfung (BAM),
(Federal Institute for Materials Research and Testing),
Richard-Willstätter-Strasse 11, D–12489 Berlin, Germany.
Tel: +49 308 104 5576   Fax: +49 308 104 5005
knut.rurack@bam.de

Specialty Keywords: Functional Dyes, Time-resolved Fluorescence, Host–guest Chemistry, Hybrid Materials.
AIM 2005 = 49.1

Development of functionalized dyes for various applications and the study of the underlying photophysical and –chemical processes (e.g. charge, electron, proton transfer). Investigation of fluorophores in confined media. Development of fluorescence lifetime standards.

-C. Trieflinger et al. (2005). Multiple switching and photogated electrochemiluminescence expressed by a dihydroazulene-boron–dipyrromethene dyad, *Angew. Chem. Int. Ed.* 44,6943.

-A.B. Descalzo et al. (2005). Rational design of a chromo- and fluorogenic hybrid chemosensor material for the detection of long-chain carboxylates, *J. Am. Chem. Soc.* 127,184.

Date submitted: 9th September 2006

Alan G. Ryder, Ph.D.

Department of Chemistry,
National University of Ireland - Galway,
Galway,
Ireland.
Tel: 353 09 149 2943   Fax: 353 09 149 4596
alan.ryder@nuigalway.ie
www.nuigalway.ie/chem/AlanR/

Specialty Keywords: Time-resolved, Petroleum, Raman.
AIM 2004 = 12.9

Senior lecturer in Chemistry leading the Nanoscale Biophotonics Laboratory, which uses fluorescence and Raman spectroscopies for the development of quantitative and qualitative analysis methods. Research areas include: metal enhanced fluorescence, lifetime based pH sensors, quantitative Raman spectroscopy, time-resolved fluorescence microscopy, and the fluorescence analysis of petroleum.

-A. G. Ryder (2005). Analysis of crude petroleum oils using fluorescence spectroscopy. *Reviews in Fluorescence 2005*, Vol. 2, 169-198.
-A.G. Ryder, S. Power, and T.J. Glynn (2003). Evaluation of acridine in Nafion as a fluorescence lifetime based pH sensor. *Appl. Spectrosc.*, 57(1), 73-79.

Date submitted: 8th September 2005

Chandran R. Sabanayagam, Ph.D.

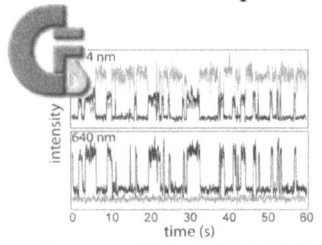

Time traces of TMR-Cy5 labelled B-DNA (R_{DA}=58Å) excited with 514nm and 640nm alternating lasers. Plots are de-interleaved data (Cy5,black; TMR,gray) showing that long-lived Cy5 dark states can cause anti-correlated donor-acceptor fluctuations.

Center for Fluorescence Spectroscopy,
Dept. of Biochemistry and Molecular Biology,
University of Maryland School of Medicine,
725 West Lombard St, Baltimore, Maryland, 21201, USA.
Tel: 410 706 2116   Fax: 410 706 8408
chandran@cfs.umbi.umd.edu
cfs.umbi.umd.edu

Specialty Keywords: Single-molecule Spectroscopy, Genomics.

My current research focuses on applying single-molecule spectroscopy to understand the structure and function of nucleic acids, and DNA / RNA polymerases. Additionally, I am interested in developing single-molecule instrumentation for applications in genomics, such as DNA sequencing and mutation detection.
-C.R. Sabanayagam, J.S. Eid and A. Meller (2005). Long timescale blinking kinetics of cyanine fluorophores conjugated to DNA and its effect on Förster resonance energy transfer. *J. Chem. Phys.* In press.
-C.R. Sabanayagam, J.S. Eid and A. Meller (2005). Using fluorescence resonance energy transfer to measure distances along individual DNA molecules: Corrections due to nonideal transfer. *J. Chem. Phys.*, 122, 061103.

Date submitted: 23rd August 2006

Kulwinder Sagoo, M.Sc.

HORIBA Jobin Yvon - IBH Limited,
45 Finnieston Street, Skypark 5,
Glasgow, G3 8JU,
United Kingdom.
Tel: +44 (0) 141 229 6789   Fax: +44 (0) 141 229 6790
Kulwinder.Sagoo@ibh.co.uk
www.jobinyvon.com

Specialty Keywords: Fluorescence lifetime spectroscopy, Time correlated single photon counting.

Member of a team providing fluorescence applications advice, support and training for all users wishing to apply single photon counting techniques to their research applications.   Have extensive academic research experience with steady-state and time-resolved fluorescence studies of biomolecule characterization for sensing applications.

-C. D. McGuinness, A.M. Macmillan, K. Sagoo, D. McLoskey, D. J. S. Birch (2006) Excitation of fluorescence decay using a 265 nm pulsed light-emitting diode: Evidence for aqueous phenylalanine rotamers. *App. Phys. Let.* 89, 063901.

Date submitted: 13th October 2006

Harekrushna Sahoo, PhD.

Chemistry Department,
International University Bremen,
Research Building III, Bremen-Schoenebeck,
Germany, D-28759,
Tel: 0049 421 200 3124   Fax: 0049 421 200 3229
hsahoo@gmail.com   &   h.sahoo@iu-bremen.de
www.iu-bremen.de/directory/02674/

Specialty Keywords: FRET, Peptide Dynamics, DBO.
AIM 2006 = 8.3

On the basis of FRET pairs with shorter Förster radius, I have employed a small fluorescent probe (2,3diazabicyclo[2.2.2]oct-2-ene, DBO) as an acceptor along with tryptophan as donor for studying the end-to-end distances in short peptides (with intervening amino acids ranging from 0 to 20). The obtained experimental result, provide an important benchmark for molecular modeling in larger proteins. The employed distribution function enables to elucidate the end-to-end distances along with the contact conformations of the polypeptides.

-H. Sahoo, D. Roccatano, M. Zacharias and W. M. Nau (2006). Distance Distributions of Short Polypeptides Recovered by Fluorescence Resonance Energy Transfer in the 10 Å Domain *J. Am. Chem. Soc.* 128 (25), 8118-8119

Date submitted: 20th June 2005

Carlota Saldanha, Ph.D.

Instituto de Bioquímica, Faculdade de Medicina de Lisboa,
Av. Prof. Egas Moniz, Lisbon,
1649-028 Lisboa,
Portugal.
Tel: +351 21 798 5136   Fax: +351 21 799 9477
saldanha@medscape.com

Specialty Keywords: Acetylcholinesterase, Membrane Fluidity, Signal Transduction.
AIM 2003 = 14.2

Enzyme kinetics studies, namely human erythrocyte and lymphocyte acetylcholinesterase, using fluorescent enzyme substrate and inhibitors. Studies of erythrocyte, lymphocyte and endothelial cells membrane fluidity and erythrocyte exovesiculation using the fluorescent probes diphenylhexatriene, trimethylamino-diphenylhexatriene and hydroxycoumarin. Studies of intracellular second messengers, namely calcium ion and nitrogen monoxide with fluorescent probes.

-C. Saldanha, N.C. Santos, J. Martins-Silva (2002) *J. Membr. Biol.*, 190, 75-82.

-C. Saldanha, N.C. Santos, J. Martins-Silva (2004) *Biochem. Mol. Biol. Educ.* 32, 250-253.

Date submitted: 7th April 2006

Nuno C. Santos, Ph.D.

Instituto Biopatologia Quimica,
Faculdade Medicina Lisboa,
Av. Prof. Egas Moniz,
1649-028 Lisbon, Portugal.
Tel: +351 21 799 9480   Fax: +351 21 799 9477
nsantos@fm.ul.pt

Specialty Keywords: Protein Intrinsic Fluorescence, Biomembranes, HIV.
AIM 2004 = 13.8

Use of steady-state and time resolved fluorescence spectroscopy (including fluorescence anisotropy, quenching, energy transfer, energy migration and red edge excitation shift) on the study of membrane proteins structure and location, intracellular ion concentration, membrane fluidity, partition of peptides and other fluorescent molecules to biomembranes, erythrocyte membrane vesiculation and binding of small fluorescent molecules to proteins. Characterization of supramolecular systems by light scattering spectroscopy.

-N.C. Santos *et al.* (2005) *Cell Biochem. Biophys.*, 43, 419-430.
-A.S. Veiga *et al.* (2005) *J. Am. Chem. Soc.*, 126, 14758-14763.

Date submitted: 1st October 2006

Suzanne F. Scarlata, Ph.D.

Dept. Physiology & Biophysics,
Stony Brook University,
Basic Science Tower, T6-146, Stony Brook,
Suffolk County, 11974-8661, USA.
Tel: 631 444 3071   Fax: 631 444 3432
Suzanne.Scarlata@sunysb.edu
www.pnb.sunysb.edu/faculty/scarlatta/scarlata.html
Specialty Keywords: Cell signalling, Protein-protein
associations, Protein-lipids interactions.
AIM 2005 = 17.8

The main focus of our laboratory is to understand the mechanism through which signals are transduced through heterotrimeric G proteins. Presently, we are focusing on the molecular basis through which G protein subunits laterally associate and activate the signalling protein, phospholipase C-beta on membrane surfaces.   In related studies, we are following the interactions and localization of these proteins in living cells using green fluorescent protein analog tags and commercial probes.

-L. Dowal, P. Provitera and S.Scarlata (2006) *J.Biol.Chem.* 281, 23999-24014.
-G. Drin, D. Douguet and S. Scarlata (2006) *Biochemistry* 45, 5712-5724.

Date submitted: 20TH July 2006

Michael Schäferling, Ph.D.

Institute of Analytical Chemistry,
Chemo - and Biosensors,
University of Regensburg,
D-93040 Regensburg.
Tel: +49 941 943 4015   Fax: +49 941 943 4064
michael.schaeferling@chemie.uni-regensburg.de
www-analytik.chemie.uni-regensburg.de/wolfbeis/institute.html
Specialty Keywords: Sensor Microarrays, Time-Resolved
Luminescence Assays, FLIM.
AIM 2004 = 12.8

Research areas: luminescent probes for enzyme activity assays, bioassays based on time-resolved FRET, sensor materials for biomedical or technical applications (fluorescent nanoparticles, oxygen and temperature sensor layers, pressure sensitive paints), fluorescence lifetime imaging (FLIM), imaging methods for microarray technology.

-Z. Lin, M. Wu, O.S. Wolfbeis, M. Schäferling: "Time-Resolved Luminescent Determination and Imaging of the Activity of Peroxidase and its Application for Enzyme-linked Immuno-sorbant Assays", *Chem. Eur. J.* 2006, *12*, 2730-2738.
-M. Schäferling, S. Nagl: "Optical Imaging Technologies for DNA- and Protein-Microarray Read Out and Quality Control"; *Anal. Bioanal.Chem.* 2006, *385*, 500-517.

Date submitted: Editor Retained.

Johannes A. Schmid, Ph.D.

Department of Vascular Biology and Thrombosis Research,
University Vienna,
Brunnerst. 59, A-1235 Vienna,
Austria.
Tel: +43 142 776 2555   Fax: +43 142 776 2550
Johannes.Schmid@univie.ac.at
www.univie.ac.at/VascBio/schmid/

Specialty Keywords: FRET, Signal Transduction.

Current research interests comprise the mechanisms of endothelial cell activation, as well as de-activation, with special focus on the signal-transduction of the NF-B pathway and its interconnection with other pathways. GFP-fusion proteins are used to elucidate the dynamics of signaling molecules in vivo. CFP and YFP-fusion proteins are used to localize protein-interactions in living cells by fluorescence resonance energy transfer microscopy.

-J.A. Schmid et al., A. Birbach, R. Hofer-Warbinek, M. Pengg, U. Burner, P.G. Furtmuller, B.R. Binder, R. de Martin R: J. Biol. Chem. 275(22), 17035-42 (2000).
-Birbach A., Gold P., Binder B.R., Hofer E., de Martin R., Schmid J.A. J. Biol. Chem. 277(13):10842-51 (2002).

Date submitted: 8th April 2006

Herbert Schneckenburger, Ph.D.

Hochschule Aalen, Biophotonics Group,
Beethovenstr. 1,
73430 Aalen,
Germany.
Tel: +49 736 156 8229   Fax: +49 736 156 8225
herbert.schneckenburger@htw-aalen.de
www.htw-aalen.de
Specialty Keywords: Biomedical Optics, Optical Microscopy,
Fluorescence Spectroscopy.
AIM 2006 = 7.0

Herbert Schneckenburger is a professor of Optics and Biophotonics, whose research is concentrated on the development and application of new methods of *in vitro* diagnostics and biomedical screening. Present techniques include fluorescence spectroscopy and microscopy, in particular time-resolved spectroscopy, total internal reflection fluorescence microscopy (TIRFM), Förster resonance energy transfer (FRET) and laser micromanipulation. Cell metabolism, membrane dynamics and photosensitization are studied within single cells and various organelles.

-H. Schneckenburger: "Total internal reflection fluorescence microscopy: technical innovations and novel applications", *Curr. Opin. Biotechnol.* 16 (2005) 13-18.

Date submitted: 17th May 2006

Bernhard Schönenberger, Ph.D.

R&D RS, Sigma-Aldrich GmbH.,
Industriestrasse 25,
CH-9471 Buchs,
Switzerland.
Tel: 0041 81 755 2609   Fax: 0041 81 755 2736
bschoene@europe.sial.com

Specialty Keywords: Amine-reactive Labels, Protein and Nucleic Acid Stains, Organic Syntheses.

Principal scientist in R&D; technology manager of center of excellence 'Detection Chemistry' of Sigma-Aldrich GmbH, Switzerland.   Recent R&D work, partly in cooperation with external groups:

Fluorescent semiconductor nanoparticles and their functionalisation for biochem. Applications.

Stains for protein and nucleic acids detection in electrophoresis.

Calibration standards for fluorescence spectroscopy.

New fluorescently labelled sec. antibodies and their applications in immunostaining.

-P. Nording, A. Rück & B. Schönenberger, BioWorld Europe 01-2006, 36

-P. Nording, A. Rück & B. Schönenberger, Laborpraxis 4-2006, Neue Fluoreszenz-Farbstoffe für den Protein-Nachweis.

Date submitted: Editor Retained.

Jörg Schroeder, Ph.D.

Institute of Physical Chemistry,
University of Göttingen,
Tammannstr. 6,
37077 Göttingen, Germany.
Tel: +49 55 139 3127   Fax: +49 55 139 3150
jschroe2@gwdg.de
www.uni-pc.gwdg.de/troe/j_schroeder/

Specialty Keywords: Photochemical Kinetics, Energy Transfer.

The main research area centers on the dynamics of elementary photoinduced reactions, in particular systematic investigations of solvation effects in supercritical fluid and liquid solution. For this purpose, time-resolved fluorescence and absorption techniques are applied to samples in environments of continuously variable density and polarity. Results are compared to classical and mixed quantum/classical non-equilibrium molecular dynamics simulations.

-J. Schroeder (2001), "Chemical Kinetics in Condensed Phases" in Encyclopedia of Chemical Physics and Physical Chemistry (eds. J.H. Moore, N.D. Spencer), Vol.I, p.711-743, IoP Publishing, Bristol, 2001.

Date submitted: 24th June 2005   **Stephen G. Schulman, Ph.D.**

College of Pharmacy,
University of Florida,
Box 100485, Gainesville,
Florida 32610-0485, USA.
Tel: 352 846 1953   Fax: 352 392 9455
Schulman@cop.ufl.edu

Specialty Keywords: Excited State Proton Transfer, pH in Aqueous-Organic Solvents, Photophysics.

Current Research Interests: Acid-base properties of organic molecules in aqueous and very concentrated aqueous-electrolyte solution. Analytical applications of exciton coupling in metal complexes and organic solids. Room temperature phosphorescence in fluid solutions. Fluorescent probes.

-R. Yang and S.G. Schulman (2003). An operational pH in Aqueous Dimethylsulfoxide based upon the acidity dependence of the rate of a simple ionic recombination reaction in the lowest excited singlet state. Talanta 60, 535-542.
-A. Fernández Gutiérrez and S.G. Schulman, eds., (2001). Fosforescencia Molecular Analítica: Una Aproximación Práctica, Universidad de Granada Press, Granada, Spain.

Date submitted: 12th October 2006   **Marcus F. Schulz, Ph.D.**

Product Manager UV-Vis & Fluorescence,
Varian Inc., Alsfelder Str. 6,
D-64289 Darmstadt,
Germany.
Tel: +49 (0) 615 217 6441   Fax: +49 (0) 615 217 6442
marcus.schulz@varianinc.com
www.varianinc.com

Specialty Keywords: Materials Science, Life Science, Solid State NMR.

As Varian's specialist for fluorescence and UV-Vis-NIR spectroscopy products in Germany, I am very interested in the whole range of fluorescence applications, which are important for life science and material science.
My background is material science with a focus on solid state NMR.

-M. Schulz, M. Tiemann, M. Fröba, C. Jäger, NMR characterization of mesostructured aluminophophates, *J. Phys. B* 104 (2000), 10473-10481.

Date submitted: Editor Retained.

Petra Schwille, Ph.D.

Experimental Biophysics,
Max-Planck-Institute for Biophysical Chemistry,
Am Fassberg 11, D-37077 Göttingen,
Germany.
Tel: +49 (0) 551 201 1165   Fax: +49 (0) 551 201 1435
pschwil@gwdg.de
www.gwdg.de/~pschwil/

Specialty Keywords: FCS, Two-Photon, Single Molecules.

Development of ultrasensitive fluorescence-based methods for detection and dynamic analysis of single or sparse biomolecules in solution, but also in the living cell. Real-time studies of fluorescent particles in open, laser-illuminated volume elements to unravel underlying inter- and intramolecular processes on time scales from nanoseconds to seconds, but also to uncover static and dynamic heterogeneities, i.e. differences in the molecular properties within ensembles of supposedly identical particles. Design of microfluidic systems for single particle manipulation.
-Heinze KG, Koltermann A, and Schwille P (2000). Simultaneous Two-Photon Excitation of Distinct Labels For Dual-Color Fluorescence Cross-Correlation Analysis. *PNAS* 97,10377-10382
Bacia K, Majoul IV, and Schwille P (2002). Probing the Endocytic Pathway in Live Cells Using Dual-Color Fluorescence Cross-Correlation Analysis. *Biophys. J.* 83,1184-1193.

Date submitted: Editor Retained.

Viviana Scognamiglio, Ph.D.

Institute of Protein Biochemistry,
Via Pietro Castellino 111,
Naples, 80131,
Italy.
Tel: +39 081 613 2312   Fax: +39 081 613 2270
scognami@dafne.ibpe.na.cnr.it

Specialty Keywords: Biosensor, Fluorescence,
Nanotechnologies.

My scientific interests deal with the development of a new generation of biosensor for analytes of high clinical, environment and food interests based on the utilization of enzymes and proteins isolated from mesophilic and thermophilic organisms. My primary goal is to identify, characterize and design enzymes and proteins to use as probes for implantable fluorescence nanodevices for the follow-up of diseases of high social impact.

Seidel, C. A. M.
Selvin, P. R.

Date submitted: Editor Retained.

Claus A. M. Seidel, Ph.D.

Heinrich-Heine-Universtaet Duesseldorf,
Institut fuer Physikalische Chemie,
Universitaetsstr, 1 Geb, 26 32 02,
40225 Duesseldorf.
Tel: +49 211 81 14755 or 15284 Fax: +49 211 81 12803
Max-Planck-Institut fuer Biophysikalische Chemie, Abt. Fuer
Spektroskkopie und Photochemische Kinetik, Am Fassberg 11,
D 37077 Goettingen.
Tel: +49 551 201 1774 Fax: +49 551 201 1501
cseidel@gwdg.de
www.mpibpc.gwdg.de/abteilungen/010/seidel/

Specialty Keywords: Single-molecule Fluorescence Spectroscopy, Multiparameter Fluorescence Detection (MFD).

It is my goal to obtain all information in a single-molecule experiment for applications in analytics and biophysics. Thus, as many fluorescence photons as possible must be detected, and a full set of fluorescence parameters must be registered by MFD: Intensity, F, lifetime, *tau* and anisotropy, r, in several spectral windows together with its time-dependence.

- R. Kuehnemuth, C. A. M. Seidel; (2001) Principles of single molecule multiparameter fluorescence spectroscopy *Single Molecules* 2, 251-254.

Date submitted: Editor Retained.

Paul R. Selvin, Ph.D.

University of Illinois at Urbana-Champaign,
Department of Physics, 363 LLP, MC704,
1110 W. Green Street, Urbana,
IL 61801-3080, USA.
Tel: (217) 244 3371 Fax: (217) 244 7187
selvin@uiuc.edu
www.physics.uiuc.edu/people/faculty/Selvin/

Specialty Keywords: Lanthanide Luminescence, FRET, Single-molecule.

We develop and use fluorescence techniques with (sub)nanometer resolution, including new forms of FRET – e.g. single-pair FRET, FRET using luminescent lanthanide chelates. A major emphasis is developing new lanthanide chelates. Applications include measuring conformational changes in myosin and ion channels.

-Cha, A., G. E. Snyder, P. R. Selvin, and F. Bezanilla. 1999. Atomic scale movement of the voltage sensing region in a potassium channel measured via spectroscopy. Nature. 402:809-813.

-Selvin, P. R. 2002. Principles and Biophysical Applications of Luminescent Lanthanide Probes. Annual Review of Biophysics and Biomolecular Structure. 31:275-302.

Date submitted: 25th October 2006

Devanand K. Shenoy, Ph.D.

Microsystems Technology Office,
Defense Advanced Research Projects Agency,
3701 N. Fairfax Drive,
Arlington, VA 22203, USA.
Tel: 571 218 4932   Fax: 703 696 2206
Devanand.shenoy@darpa.mil
www.darpa.mil

Specialty Keywords: Photonics, Correlation Spectroscopy,
Single Molecule Detection.

Dr. Shenoy is currently a program manager at the Microsystems Technology Office at DARPA where he is managing a program on Supermolecular photonics and is developing a new program in the area of standoff detection of explosives using optical spectroscopic technologies. Dr. Shenoy's recent interests have also included the areas of supramolecules such as cavitands and liquid crystals for chemical sensing and stochastic sensing of biological molecules using synthetic nanopores.

Date submitted: 29th September 2006

Claudio H. Sibata, Ph.D.

Radiation Oncology,
ECU School of Medicine,
600 Moye Blvd, Greenville,
Pitt County, 27834, USA.
Phone: (252) 744-2900 Fax: (252) 744-3780
sibatac@ ecu.edu
www.ecu.edu/radiationoncology
Specialty Keywords: Cancer, Photodynamic therapy, Optical biopsy.
AIM 2005 = 11.5

Research interests are optical biopsy, photodynamic therapy optimization for oncology patients, refine and improve therapy both by clinical modifications and dosimetry enhancement.

-RR Allison, VS Bagnato, R Cuenca, GH Downie and CH Sibata (2006). The future of photodynamic therapy in oncology. Future Oncol. 2(1), 53-716.

-RR Allison, GH Downie, R Cuenca, CH Sibata (2006). A brief review of photodynamic therapy in oncology. Amer. J of Oncology Review. 5(2), 111-117.

Siebert, R.
Siemiarczuk, A.

Date submitted: 23rd August 2004

Reiner Siebert, M.D.

Institute of Human Genetics, University Hospital Kiel,
Schwanenweg 24, Kiel,
Schleswig-Holstein, 24105,
Germany.
Tel: +49 431 597 1779   Fax: +49 431 597 1880
rsiebert@medgen.uni-kiel.de

Specialty Keywords: Combined Immunofluorescence and Fluorescence in Situ Hybridization (FICTION).

With regard to fluorescence microscopy, my research focuses on the development of fluorescence in situ hybridization (FISH) assays for the detection of chromosomal abnormalities in tumors, as well as on the technical improvements of the fluorescence immunophenotyping and interphase cytogenetics (FICTION) technique. The next goal of my research team is to optimize an automated platform for the detection of rare tumor cells and spot counting of multicolor hybridization signals.

-J.I. Martin-Subero, I. Chudoba, L. Harder, S. Gesk, W. Grote, F.J. Novo, M.J. Calasanz, R. Siebert (2002). Multicolor-FICTION: Expanding the Possibilities of Combined Morphologic, Immunophenotypic, and Genetic Single Cell Analyses, Am. J. Pathol., 161, 413-420.

Date submitted: 22nd June 2006

Aleksander Siemiarczuk, Ph.D.

Photon Technology International,
347 Consortium Court, London,
Ontario, N6E 2S8,
Canada.
Tel: 519 668 6920
asiemiarczuk@pti-can.com
www.pti-nj.com

Specialty Keywords: Time Resolved Fluorescence, TICT, Protein Fluorescence, Lifetime Distributions.

My activities include: development of stroboscopic time-resolved instrumentation and research in molecular photophysics, specifically: time-resolved fluorescence of proteins, lifetime distributions, complexes with cyclodextrins, intramolecular and solvation dynamics, new methodology to study polydispersity of micelles, long-range electron transfer in linked porphyrin-quinone derivatives, co-discovery of Twisted Intramolecular Charge Transfer States (TICT).

-Paul A. Jelliss, Keith M. Wampler, and Aleksander Siemiarczuk (2005) Enhanced Dual Visible Light Fluorescence from the 2,2'-Dipyridyl Tungsten Alkylidyne Complex [W($\equiv$CC6H4NMe2-4)(O2CCF3)(CO)2{κ2-2,2'-(NC5H4)2}]: An Organometallic Twisted Intramolecular Charge Transfer State, Organometallics 24(4), 707 – 714

Date submitted: 17th October 2006

Ewa Sikorska, Ph.D.

Faculty of Commodity Science,
The Poznań University of Economics,
al. Niepodleglosci 10,
Poznan, 60-967, Poland.
Tel: +48 61 856 9040   Fax: +48 61 854 3993
ewa.sikorska@ae.poznan.pl
www.kai.ae.poznan.pl

Specialty Keywords: Fluorescence, Flavins, Chemometrics.
AIM 2005 = 20.0

Study of photophysics and photochemistry of biologically important molecules (flavins, isoalloxazines and alloxazines); study of excited-state double proton-transfer in alloxazines in solution and in crystal solids; application of spectroscopic methods and chemometrics in analysis of complex material e.g.: foodstuff, kraft pulps, dental materials.

-E. Sikorska, T. Gorecki, I. V. Khmelinskii, M. Sikorski and D. De Keukeleire (2006) Monitoring beer during storage by fluorescence spectroscopy, *Food Chem.,* 96, 632-639.
-E. Sikorska, I. V. Khmelinskii, M. Kubicki, W. Prukala, G. Nowacka, A. Siemiarczuk, J. Koput, L. F. V. Ferreira and M. Sikorski (2005) Hydrogen-bonded complexes of lumichrome, *J. Phys. Chem. A,* 109, 1785-1794.

Date submitted: 30th October 2006

Manoj K. Singh, Ph.D.

Spectroscopy Division, BARC,
Trombay, Mumbai,
400 085,
India.
Tel: +91 222 559 0351
mksingh@barc.gov.in
L

Specialty Keywords: Photophysics, Photochemistry, Single Molecule Fluorescence.

The main area of our research has so far been the photophysics and photochemistry of probes used in biological research. Recently, we have been involved in the development of facility for single molecule fluorescence spectroscopy and its application to investigate macro-molecules.

-M. K. Singh, H. Pal, A. S. R. Koti and A. V. Sapre (2004). Photophysical properties and rotational relaxation dynamics of neutral red bound to β–cyclodextrin *J. Phys. Chem. A* 108, 1465-1474.

-K. D. Osborn, M. K. Singh, R. J. B. Urbauer and C. K. Johnson (2003). Maximum-Likelihood approach to single molecule polarization modulation analysis *CHEMPHYSCHEM* 4, 1005-1011.

Date submitted: 17th October 2006   **Ramendra Kumar Singh, Ph.D.**

Nucleic Acids Research Laboratory, Department of,
Chemistry, University of Allahabad,
Allahabad, 211002,
India.
Tel: / Fax: +91 532 246 1005
singhramk@rediffmail.com   &   singhramenk@yahoo.com

Specialty Keywords: DNA / RNA, Therapeutics, Diagnostics.
AIM 2006 = 4.9

The research interest involves the development of nucleosides, nucleotides and oligonucleotides as therapeutics. The nucleosides analogues are developed as inhibitors of nucleic acids metabolic processes like replication, transcription or reverse transcription by chain termination methods or blocking the capping process of viral mRNA and thus interfering at the level of translation. The other part of major research activities is development of fluorescently labelled oligonucleotides to be used as diagnostics or in molecular biology.

References
-S. Singh & R. K. Singh (2006), *Current Science,* 25 (6), 836.
-S. Sinha, R. Srivastava, E. De Clercq & R. K. Singh (2004), *Nucleosides, Nucleotides and Nucleic Acids*, 23(12), 1815.

Date submitted: 1st August 2004   **Harald H. Sitte, M.D.**

Medical University Vienna,
Institute of Pharmacology,
Währingerstr. 13a, A-1090 Vienna,
Austria.
Tel: +43 1 42 776 4123   Fax: +43 1 42 776 4122
harald.sitte@meduniwien.ac.at

Specialty Keywords: Fluorescence Microscopy, Fluorescence Resonance Energy Transfer, Membrane Proteins.
AIM 2003 = 38.8

My research focuses on the understanding of the quaternary structure of membrane proteins, i.e. transport proteins like the serotonin or the GABA transporter. We use Fluorescence Resonance Energy Transfer Microscopy to learn more about their structural constraints and the impact, oligomerization may have on the function of these proteins.

-Farhan H, Korkhov VM, Paulitschke V, Dorostakar MM, Schoze P, Kudlacek O, Freissmuth M, Sitte HH. (2004) Two discontinuous segments in the carboxyl terminus are required for membrane targeting of the rat gamma-aminobutyric acid transporter-1 (GAT1) J Biol Chem. 2004; 279:28553-63.
-Sitte HH, Farhan H, Javitch JA (2004) Oligomerization as a determinant of transporter function and tracking molecular Interventions 4 (1): 38-47.

Date submitted: 12th October 2006 **Clint B. Smith, Ph.D.**

U.S. Army Engineer Research and Development Center,
Fluorescence Spectroscopy Laboratory,
7701 Telegraph Road, Alexandria,
VA 22315, USA.
Tel: 703 428 8203   Fax: 703 428 8176
clint.b.smith@erdc.usace.army.mil
www.tec.army.mil
Specialty Keywords: Fluorescence Spectroscopy, Quantum
Dots, Waterborne Pathogens, Silicate Nanoparticles.
AIM 2006 = 7.2

Research in the laboratory involves fluorescence remote sensing applications for biological, chemical, and radiological threats.   Novel fluorescent probes and molecularly imprinted polymers are utilized for the detection of pathogens existing in waterways and soils using state-of-the-art fluorescent spectrometers.   Applications are geared toward the imaging domain and will be developed after performing successful laboratory experiments binding select probes to specific targets.

-Smith, C.*, Balaji, T., Anderson, J. and Tepper, G. (2006). Optical Bar Code Recognition of Methyl Salicylate (MES) for Environmental Monitoring using Fluorescence Resonance Energy Transfer (FRET) on Thin Films. SPIE Optics East. In press.

Date submitted: 11th April 2006 **Trevor A. Smith, Ph.D.**

School of Chemistry,
University of Melbourne,
Victoria, 3010,
Australia.
Tel: +61 3 8 344 6272   Fax: +61 3 9 347 5180
trevoras@unimelb.edu.au
www.chemistry.unimelb.edu.au/people/smith.php

Specialty Keywords: Time-resolved fluorescence, Anisotropy,
Microscopy.

Research Interests: Ultrafast laser spectroscopic techniques applied to photophysical processes in macromolecules.   Time-resolved fluorescence and advanced optical microscopy techniques including multi-photon & confocal fluorescence microscopy, single molecule spectroscopy and fluorescence correlation methods.   Time-resolved fluorescence anisotropy measurements, time-resolved evanescent wave-induced fluorescence techniques.
-T.A. Smith, C.A. Scholes and M.L. Gee, (2005) "Polarized Time-Resolved Evanescent Wave-Induced Fluorescence Measurements" Reviews in Fluorescence, Springer Science, pp 245-270,
-D.E. Gómez, J. van Embden, J. Jasieniak, T.A. Smith and P. Mulvaney, (2006) "Blinking and Surface Chemistry of Single CdSe Nanocrystals", Small, 2(2), 204-208

Date submitted: Editor Retained.

Peter T. C. So, Ph.D.

Department of Mechanical Engineering,
Department of Biological Engineering,
Massachusetts Institute of Technology,
3-461, 77 Mass Ave,
Cambridge, MA, 02139,
USA.
Tel: (617) 253 6552   Fax: (617) 258 9346
ptso@mit.edu
Specialty Keywords: Multi-photon Microscopy, Time-resolved
Spectroscopy, Correlation Spectroscopy.

My research focuses on the development of instrumentation for biomedical studies. Recent projects in my laboratory include video rate two-photon microscopy, fluorescence correlation spectroscopy, 3-D image cytometery. These instruments are applied in studies such as: Protein dynamics, cellular mechanotransduction, tissue carcinogenesis, and non-invasive optical biopsy.

-So et al., "Two-Photon Excitation Microscopy", Annu. Rev. Biomed. Eng., 2, 399-429 (2001). Huang et al., "Three-Dimensional Cellular Deformation Analysis with a Two-Photon Magnetic Manipulator Workstation," Biophys. J., 82, 2211-2223 (2002).

Date submitted: 15th August 2005

Steven A. Soper, Ph.D.

Department of Chemistry,
Louisiana State University,
232 Choppin Hall, Baton Rouge,
LA 70803, USA.
Tel: 225 578 1527   Fax: 225 578 3458
chsope@lsu.edu
www.chmm.lsu.edu
Specialty Keywords: Near-IR Fluorescence, Microfluidics,
Single Molecule Detection.
AIM 2004 = 32.7

Our work is focused on developing novel fluorescence-based strategies for analyzing biological material, in particular genes that harbor mutations and possess high diagnostic value for certain cancers. These projects use microfluidics for sample processing and single molecule fluorescence in the near-IR to alleviate background artifacts arising from sample matrix interferences. Our work has developed assays for screening point mutations that carry high diagnostic value for colorectal cancers.

-Single Pair FRET Generated from Molecular Beacon Probes for the Real-Time Analysis of Low Abundant Mutations, M. Wabuyele, W. Stryjewski, Y.-Wei Cheng, F. Barany, H. Farquar, R.P. Hammer and S.A. Soper, *JACS* 125 (2003) 6937-6945.

Date submitted: Editor Retained. **Ian Soutar, Ph.D.**

Chemistry Department,
University of Sheffield,
Brook Hill, Sheffield,
S3 7HF, UK.
Tel: +44 (0)114 222 9561   Fax: +44 (0)114 273 8673
i.soutar@sheffield.ac.uk

Specialty Keywords: Anisotropy, Energy Harvesting, Polymers.

Research Interests: Studies of polymer behavior both in solution and the solid state using time-resolved emission anisotropy, Water-soluble polymers, Smart systems, Polymers for energy harvesting and solar energy conversion.

-D. Allsop, L. Swanson, I. Soutar et al. (2001) "Fluorescence Anisotropy: A Method for Early Detection of Alzheimer β-Peptide Aggregation" *Biochem. Biophys. Res. Comm.,* 285, 58-63.
-C.K. Chee, I. Soutar et al. (2001) "Time-resolved Fluorescence Studies of the Interactions Between the Thermoresponsive host, PNIPAM, and Pyrene" *Polymer,* 42, 1067-1071.

Date submitted: 8ᵗʰ April 2006 **C. Michael Stanley, Ph.D.**

Chroma Technology Corporation,
10 Imtec Lane,
Rockingham, Vermont, 05101,
USA.
Tel: 802 428 2500 / 800 824 7662   Fax: 802 428 2525
m@chroma.com
www.chroma.com

Specialty Keywords: Confocal, Multi-Photon, Laser Based Applications.

My previous research and experience, in both confocal and widefield imaging systems, allows me to collaborate, design, and trouble-shoot fluorescent experimental designs. Current areas of emphasis are laser-based systems, in both one and multi-photon applications.

Chroma Technology's filters have been developed for a variety of applications: low-light microscopy, cytometry; spectroscopy and laser-based confocal and multi-photon instrumentation.

Date submitted: 21st August 2006

Elias Stathatos, Ph.D.

Engineering Science Department,
University of Patras,
26500 Patras,
Greece.
Tel: +30 261 099 7587   Fax: +30 261 099 7803
stathatos@des.upatras.gr

Specialty Keywords: Photophysics, Sol-gel, Solid-state
Electrolytes.
AIM 2006 = 17.3

Research interests include steady-state and time-resolved fluorescence characterization of nanocomposite thin films and transparent solid matrices. Applications involve dye-sensitized photoelectrochemical solar cells, photocatalytic metal oxide surfaces, lasing in nanocomposite and organic materials and electroluminescence of ligand lanthanide complexes.

-Elias Stathatos, Lefkia Panayiotidou, Panagiotis Lianos and Anastasios D. Keramidas. *Thin Solid Films,* 496, (2006), 489-493.

- Elias Stathatos and Panagiotis Lianos *Chemical Physics Letters*, 417, (2006), 406-409.

Date submitted: 6th April 2006

Lorenzo Stella, Ph.D.

Department of Chemical Sciences and Technologies,
University of Roma Tor Vergata,
Via della ricerca scientifica,
00133, Roma, Italy.
Tel: +39 067 259 4463   Fax: +39 067 259 4328
stella@stc.uniroma2.it
www.stc.uniroma2.it/physchem/
Specialty Keywords: Peptide and Protein Structure and
Dynamics, Peptide-membrane Interactions.
AIM 2006 = 35.5

My main research focus is the application of fluorescence spectroscopy in the study of protein and peptide structure, function and dynamics. Some current research projects: mechanism of action of antimicrobial peptides and their interaction with membranes; design and characterization of peptide-based molecular devices utilizing photophysical processes for memories, switches and energy conversion; comparisons between time-resolved fluorescence spectroscopy data and molecular dynamics simulations.

-M. Venanzi et al. (2006) Dynamics of formation of a helix-turn-helix structure in a membrane-active peptide: a time-resolved spectroscopic study. *ChemBioChem* 7, 43-45.

Date submitted: 4th[th] April 2006

Charles Neal Stewart, Jr., Ph.D.

Department of Plant Sciences,
University of Tennessee,
252 Ellington PSB,
Knoxville, TN 37996, USA.
Tel: 001 865 974 6487   Fax: 001 865 946 1989
nealstewart@utk.edu
plantsciences.utk.edu/stewart.htm

Specialty Keywords: GMOs, Fluorescent Proteins, Plants.
AIM 2005 = 58.4

Research is performed towards the development of fluorescent transgenic plants for various purposes. Plants expressing GFP or other fluorescent proteins can be used to mark other transgenic products for marking GMOs in vivo. In addition, work is being performed to use FP-transgenic plants as environmental sensors—*phytosensors*—that can be used to detect and report the presence of explosives and other hazardous chemicals as well as biologicals such as pathogens.

-Stewart, C.N., Jr. 2005. Monitoring the presence and expression of transgenes in living plants. Trends in Plant Science 10:390-396.

Date submitted: Editor Retained.

Daniel W. Stockholm, Ph.D.

Laboratoire d'imagerie, Genethon,
1 bis rue de l'Internationale,
Evry, 91000,
France.
Tel: 01 60 77 8698
stockho@genethon.fr

Specialty Keywords: Confocal Microscopy, Muscle Visualization, Real-time PCR.

We are part of a research center focussed on gene therapy and run a core service for imaging with 2 confocal microscopes including a multiphon. We use FRET for the study of calpain function and are developing techniques for the intra vital imaging. We also acquired some expertise in real-time PCR and use it extensively for gene expression studies and viral titration.

-Stockholm D, et al. , *Am J Physiol Cell Physiol,* 2001, Jun;280(6):C1561-9.

-Feasson L, Stockholm D, et al. *J Physiol.* 2002 Aug 15;543(Pt 1):297-306.

Date submitted: 4th September 2006

Karel W. J. Stoop, M.Sc.

Lambert Instruments,
Turfweg 4,
9313 TH Leutingewolde,
The Netherlands.
Tel: +31 50 501 8461   Fax: +31 50 501 0034
kstoop@lambert-instruments.com
www.lambert-instruments.com

Specialty Keywords: Fluorescence Lifetime Imaging
Microscopy, Image Intensifiers, FRET

My work is focused on the ongoing development of the Fluorescence Lifetime Imaging Microscopy (FLIM), on a wide field microscope. We work in the frequency domain. Our specialty is the use of LED`s as the modulated light source, rather then (expensive) lasers. The FLIM-system is mostly used for FRET of the protein pairs GFP-RFP and CFP-YFP.

-L.K. van Geest, K.W.J. Stoop (2003). FLIM on a wide field fluorescence microscope, *Published in "Letters in Peptide Science" Issue: Volume 10, Numbers 5-6 2003, Pages: 501 – 510.*
-K.W.J. Stoop, K. Jalink, S.J.G. de Jong, L.K. van Geest (2004) Measuring FRET in living cells with FLIM *Proceedings of 8th Chinese International Peptide Symposium.*

Date submitted: 19th October 2006

Robert M. Strongin, Ph.D.

Department of Chemistry, Louisiana State University,
232 Choppin Hall, Baton Rouge,
Louisiana, 70803,
USA.
Tel: 225 578 3238   Fax: 225 578 3458
rstrong@lsu.edu
www.chem.lsu.edu/htdocs/people/rmstrongin/
Specialty Keywords: Supramolecular Chemistry, Molecular
Recognition, Sensors.
AIM 2006 = 58.8

Our research group investigates the design and synthesis of organic receptors and the modulation of their colorimetric and fluorimetric properties. The selective determination of saccharides, amino acids and related molecules in biological media is a major focus of our program.

-Jiang, S.; Escobedo, J. O.; Kim, K. K.; Alpturk, O.; Samoie, G.; Fakayode, S. O.; Rusin, O.; Warner, I. M.; Strongin, R. M. "Stereochemical and Regiochemical Trends in the Selective Spectrophotometric Detection of Saccharides," *J. Am. Chem. Soc.* 2006, *128*, 12221-12228.
-Alptürk, O.; Rusin, O.; Fakayode, S. O.; Wang, W.; Escobedo, J. O.; Warner, I. M.; Král, V.; Strongin, R. M. "Lanthanide Complexes as Fluorescent Indicators for Neutral Sugars and Small Molecule Cancer Biomarkers," *Proc. Natl. Acad. Sci. USA* 2006, *103*, 9756-9760.

Date submitted: 11th August 2005

Klaus Suhling, Ph.D.

Department of Physics,
King's College London,
Strand, London, WC2R 2LS,
UK.
Tel: +44 (0)207 848 2119   Fax: +44 (0)207 848 2420
klaus.suhling@kcl.ac.uk
www.kcl.ac.uk/kis/schools/phys_eng/physics/people/suhling.ht
ml

Specialty Keywords: Time-correlated Single Photon Counting,
Time-resolved Fluorescence, Fluorescence Lifetime Imaging.

Fluorescence imaging techniques are minimally invasive and can be applied to living cells. The fluorescence emission can be characterized not only by its intensity and position, by also by its fluorescence lifetime, polarization and wavelength. Each of these observables provides an additional dimension which contains information about the biophysical environment of specific proteins.

-B.Treanor, P.M.P. Lanigan, K. Suhling, T.Schreiber, I. Munro, M.A.A. Neil, D. Phillips, D.M. Davies, and P.M.W. French. Imaging fluorescence lifetime hetrogeneity applied to GFP-tagged MHC protein at an immunological synapse. J Micros 217 (1) 36-43, 2005.
-K.Suhling, P.M.W. French and D.Phillips. Time-resolved fluorescence microscopy. Photochemical and Photobiological Sciences 4, 13-22, 2005.

Date submitted: 6th April 2006

Johanna M. Suomi, Ph.D.

Laboratory of Inorganic and Analytical Chemistry,
Helsinki University of Technology,
P.O. Box 6100, Espoo,
FIN-02015 TKK, Finland.
Tel: +358 9 451 2593   Fax: +358 946 2373
johanna.suomi@tkk.fi

Specialty Keywords: Capillary Electrophoresis, Sample
Pretreatment, Electrogenerated Chemiluminescence.
AIM 2005 = 15.3

Johanna Suomi did her Ph.D. research on capillary electrophoretic analysis of monoterpenoids in natural samples. After finishing her Ph.D. degree she has concentrated on electrogenerated chemiluminescence studies as a member of Professor Kulmala's hot electron ECL research group. Johanna Suomi is currently employed as senior lecturer at Helsinki University of Technology. She also holds title of docent at University of Helsinki.

-J. Suomi, T. Ylinen, M. Håkansson, M. Helin, Q. Jiang, T. Ala-Kleme, S. Kulmala, (2006), Hot electron-induced electrochemiluminescence of fluorescein in aqueous solution, J. Electroanal. Chem. 586 (1), 49-55.

Date submitted: Editor Retained.

John C. Sutherland, Ph.D.

Biology Department,
Brookhaven National Laboratory,
Upton, NY, 11973,
USA.
Tel: 631 344 3279
jcs@bnl.gov
bnlstb.bio.bnl.gov/biodocs/structure/J_Sutherland.htmlx

Specialty Keywords: Time-resolved Fluorescence and Circular Dichroism using Synchrotron Radiation, DNA Damage Quantitation by Gel Electrophoresis and Single Molecule Sizing.

Pioneered the use of synchrotron radiation for the measurement of circular dichroism and time-resolved fluorescence spectroscopy in the ultraviolet/visible spectral regions. Invented the Fluorescence Omnylizer, a single-photon counting detector that records the time-delay, wavelength and polarization of each detected photon. First to use CCD camera to record image of DNA fluorescence in electrophoretic gels. Uses gel fluorescence or single-molecule laser fluorescence sizing to quantify DNA damage by average length analysis.

Date submitted: 1st October 2006

Meenakshisundaram Swaminathan, Ph.D.

Department of Chemistry, Annamalai University,
Annamalai Nagar 608 002,
Chidambaram, Tamil nadu,
India.
Tel: 91 414 422 0572
chemsam@yahoo.com

Specialty Keywords: Inclusion complex, Photoprotropism, cyclodextrin.
AIM 2006 = 25.8

Our Research is mainly focused on the spectral and prototropic behavior of organic fluorophores in aqueous and beta cyclodextrin media using fluorimetric technique. In addition florescence quenching studies are carried out with halomethanes, metal ions and anions as quenchers. A green technology is developed for the degradation of toxic chemicals using newly synthesized solar photocatalysts.

-I.M.V. Enoch and M.Swaminathan (2006).J.Fluoresc.16(4), 501-510.

-I.M.V. Enoch and M.Swaminathan (2006) J.Fluoresc.16(5) 697-704.

Date submitted: 13th September 2005

Linda Swanson, Ph.D.

Chemistry Department,
University of Sheffield,
Brook Hill, Sheffield,
S3 7HF, UK.
Tel: +44 (0)114 222 9564   Fax: +44 (0)114 273 8673
l.swanson@sheffield.ac.uk

Specialty Keywords: Anisotropy, Smart Polymers, Polymer Dynamics.

Research Interests: Anisotropy studies of the conformational behavior of smart polymers. Polymer dynamics. Polymer interactions (synthetic and biomacromolecules). Polymer relaxation behavior in the solid state. Novel polymeric materials for enhanced solar energy conversion.

-L. Swanson, et al., (2001). "Manipulating the thermoresponsive behaviorof PNIPAM "*Macromolecules* 34, 544-754.
-N. J. Flint, S. Gardebrecht and L. Swanson, (1998). "Luminescence investigations of smart microgel systems", *J. Fluorescence,* 8, 343-353.

Date submitted: 5th April 2006

Kerry M. Swift, M.S.

Abbott Laboratories,
Global Pharmaceutical Research and Development,
Department of Structural Biology R46Y / AP9LL,
Abbott Park, IL 60064-6114, USA.
Tel: 847 937 7289   Fax: 847 935 0143
Kerry.Swift@abbott.com

Specialty Keywords: Drug Discovery, Binding, Fluorescence Lifetimes, FP, FCS, HTS.

One of the research goals of the optical spectroscopy group is to characterize or improve fluorescent probe-based assays for testing of drug-like compounds. Furthermore, we sometimes use the intrinsic fluorescence of target proteins to study their structure or binding. We are also developing the use of Raman & UV fluorescence microscopy on protein crystals.

-Russell A. Judge, Kerry Swift and Carlos González (2005). An ultraviolet fluorescence-based method for identifying and distinguishing protein crystals *Acta Crystallographica* D61, 60–66.
-Sergey Y. Tetin, Kerry M. Swift and Edmund D. Matayoshi (2002). Measuring antibody affinity and performing immunoassay at the single molecule level *Analytical Biochemistry* 307(1) 84-91.

Date submitted: 11th October 2006

Henryk Szmacinski, Ph.D.

Center for Fluorescence Spectroscopy,
Dept. of Biochemistry and Molecular Biology,
University of Maryland School of Medicine,
725 West Lombard St, Baltimore, Maryland, 21201, USA.
Tel: 410 706 7500   Fax: 410 706 8408
henry@cfs.umbi.umd.edu

Specialty Keywords: Spectroscopy, Fluorescence Probes, Optical Sensing.

My research interests include UV/VIS spectroscopy, optical sensors and biosensors, frequency-domain time resolved spectroscopy, and multi-photon microscopy. This involves of application of fluorescence lifetime to chemical sensing and imaging, immunoassays, DNA hybridization and cellular studies. Current interest is in development of disposable sensor arrays for biotechnology and clinical chemistry and exploring enhanced fluorescence using metallic nano-structures.

-T.D. Corrigan, S. Guo, Szmacinski H., and J. Phaneuf (2006). Systematic study of the size and spacing dependence of Ag nanoparticle enhanced fluorescence using electron-beam lithography. *Appl. Phys. Lett.* 88:101112.

Date submitted: 28th June 2004

Fumio Tanaka, Ph.D.

Mie Prefectural College of Nursing,
Yumegaoka 1-1-1,
Tsu 514 0116,
Japan.
Tel / Fax: 81 59 233 5640
fumio.tanaka@mcn.ac.jp

Specialty Keywords: Flavin, Flavoprotein, Time-resolved Fluorescence, Theoretical Analysis.

I am working mostly on the time-resolved fluorescence of tryptophan and flavins in proteins in sub-picosecond region.   I was much inspired on theory of anisotropy by knowing Weber's the Additivity Law of Polarization.   I still have interested in developing the theory of fluorescence anisotropy.

-Choswojan, H., Taniguchi, S., Mataga, N., Tanaka, F., Visser, A. J. W. G.,2003, The stacked flavin adenine dinucleotide conformation in water is fluorescent on picosecond timescale, Chem. Phys. Lett., 378, 354-358.
-Tanaka, F., 2004, Theoretical analyses of time-resolved fluorescence in simple and biological systems, Recent Research Developments in Physical Chemistry, Vol 7, 1-41.

Date submitted: Editor Retained.

Hans J. Tanke, Ph.D.

Department Molecular Cell Biology,
Leiden University Medical Center,
Wassenaarseweg 72, 2333 AL Leiden,
The Netherlands.
Tel: +31 71 527 6196   Fax: +31 71 527 6180
H.J.Tanke@lumc.nl

Specialty Keywords: Fluorescence Technology, Molecular Analysis, Microscopy.

The study of the molecular composition of cells and chromosomes, using fluorescence labeling technology (FISH, immunocytochemistry, GFP) and (automated) digital microscopy, in order to unravel the molecular mechanisms that determine normal and abnormal cell function. The use of this information and methodology to develop improved diagnostic methods to be applied in the field of genetics, haematology and oncology.

-Rijke F.v.d. et al. Up-converting phoshor reporters for nucleic acid microarrays. Nature Biotechnology 19:273-276, 2001. Ref. 2: Snaar SP et al. Mutational analysis of fibrillarin and its mobility in living cells. J. Cell Biol. 151:653-662, 2000.

Date submitted: 28th June 2005

Olga Nikolaevna. Tchaikovskaya, Ph.D.

Photonics Department,
Siberian Physical Technical Institute at Tomsk State University,
1, Novo-Sobornaya,
Tomsk, 634050, Russia.
Tel: +7 382 253 3426   Fax: +7 382 253 3034
tchon@phys.tsu.ru

Specialty Keywords: Photochemistry, Fluorescent Spectroscopy, Photolysis.

The method of fluorescent spectroscopy is used to investigate the influence of the exciting radiation wavelength on phototransformations of phenols and natural biomaterials in water under UV irradiation (222, 266, 283, 308 nm).

-P.Plaza, M.Mahet, O.N. Tchaikovskaya, and M.Martin/ Excitation energy effect on the early photophysics of hypericin in solution/ ChemPhysLetter, 2005, V.408. P.96-100.

-O.N. Tchaikovskaya, I. V. Sokolova, N.V. Udina/Fluorescence analysis of photoinduced degradation of ecotoxicants in presence humic acids/ J.Biological and Chmical Luminescence, 2005,V.20,P.101-107.

Date submitted: 12th October 2006

Jamshid P. Temirov, Ph.D.

Center for Integrated Nanotechnologies (CINT),
Los Alamos National Laboratory, MS G755,
Los Alamos, NM 87545,
USA.
Tel: 505 665 0151   Fax: 505 665 9427
jama@lanl.gov

Specialty Keywords: Single-molecule, Spectroscopy, Imaging.
AIM 2004 = 6.2

Dr. Temirov's current research is aimed at the study of interactions between individual antibodies with their antigens with the goal of developing highly quantitative single molecule antibody arrays for multiplexed target detection. Also, Dr. Temirov characterizes new fluorophores (proteins and nanoparticles) using a wide variety of Single Molecule fluorescence techniques including, wide field TIR imaging, Fluorescence Correlation Spectroscopy, Time-Correlated Single Photon Counting and photon antibunching.

-J. Temirov, A. Bradbury, J. Werner (2006) *Progress in Biomedical Optics and Imaging – Proc. of SPIE.* v. 6092.

-M. Dai, H. Fisher, J. Temirov, C. Kiss, L. Phipps, J. Werner, A. Bradbury (2006) *PEDS,* in press.

Date submitted: 29th June 2004

Richard B. Thompson, Ph.D.

Department of Biochemistry and Molecular Biology,
University of Maryland School of Medicine,
108 N. Greene Street, Baltimore,
Maryland 21201, USA.
Tel: 410 706 7142   Fax: 410 706 7122
rthompso@umaryland.edu

Specialty Keywords: Biosensors, Fiber Optic Sensors, Metal Ions

Our work focuses on fluorescence-based biosensors and fiber optic biosensors. Our metal ion biosensors employ carbonic anhydrase II variants as recognition molecules, which transduce the concentrations of metal ions (Cu(II), Zn(II), and others) as changes in fluorescence lifetime, polarization, or intensity ratio. Carbonic anhydrase gives the sensor unmatched sensitivity (picomolar) and selectivity (demonstrated in sea water and cerebrospinal fluid), which both can be modulated by subtle changes in the protein structure. Optical fiber sensors permit continuous monitoring in situ.

-C. A. Fierke and R. B. Thompson (2001). Fluorescence-based biosensing of zinc using carbonic anhydrase. BioMetals 14 (3-4), 205-222.

-H. H. Zeng, et al. (2003). Real-time determination of picomolar free Cu(II) in seawater using a fluorescence-based fiber optic biosensor. Anal. Chem. 75, 6807-6812.

Date submitted: 6th April 2006 **Leann Tilley, Ph.D.**

Department of Biochemistry,
La Trobe University,
Melbourne, Victoria, 3086,
Australia.
Tel: 61 39 479 1375   Fax: 61 39 479 2467
l.tilley@latrobe.edu.au

Specialty Keywords: FRAP, Confocal Microscopy, GFP.
AIM 2005 = 38.4

Prof Tilley uses fluorescence techniques to study malaria parasite-infected erythrocytes. She has employed a series of optical spectroscopy techniques including UV-Vis spectroscopy, fluorescence polarisation and time-resolved phosphorescence anisotropy. She has developed protocols for incorporating exogenous fluorescent probes into parasite-infected erythrocytes and has used molecular biology protocols to produce transfectants expressing green fluorescent protein chimeras. She has set up a facility for quantitative measurements of fluorescence recovery after photobleaching and fluorescence correlation spectroscopy using the confocal microscope.

-Klonis N, Rug M, Wickham M, Harper I, Cowman A and Tilley L (2002) Fluorescence photobleaching analysis for the study of cellular dynamics. Eur J Biophys (review), 31: 36-51.

Date submitted: 11th August 2005 **Alphonse Tine, Ph.D.**

Faculté des Sciences et Techniques,
Université Cheikh Anta DIOP,
Dakar,
Senegal.
Tel: +221 824 6318
alphtine@ucad.sn
www.ucad.sn

Specialty Keywords: Photochemistry, Photophysics,
Fluorescence Sensing.

My research is focused on developing new fluorescence sensing methods of histamine and harmful metals (lead, arsenic, mercuric and cadmium) in halieutic products. My laboratory also developed new sensitive fluorimetric methods for monitoring pesticide residues in environmental matrices. After a large ground and theoretical researches on heterocyclic compounds like coumarine and indole derivatives; we applied these compounds for agricultural interest like regulator of plant growth (groundnut and haricot bean).

-A. Tine *et al* (2003). Talanta 60, 581-590;
-A. Tine *et al* (2005). JOFL 91 in press
-A. Tine *et al* (2003). Talanta 60, 571-579.

Date submitted: 1st August 2006

Dmitri Toptygin, Ph.D.

Johns Hopkins University,
Department of Biology,
3400 North Charles Street,
Baltimore, Maryland 21218, USA.
Tel: 410 516 7300   Fax: 410 516 5213
toptygin@jhu.edu

Specialty Keywords: Quantum Radiophysics, Fluorescence,
Data Analysis.
AIM 2006 = 7.4

Theory: the fundamental physics laws that govern the absorption and emission of photons by fluorescent molecules in solutions, in liquid crystals, near interfaces, near or inside nanoparticles, and near other fluorescent molecules. Experiment: elimination of systematic errors in time-resolved fluorescence instrumentation, both time-correlated photon counting and frequency domain. Data analysis: efficient □2 minimizatoin with hundreds of fitting parameters.

-D. Toptygin, R. S. Savtchenko, N. D. Meadow, S. Roseman, L. Brand (2002) *J. Phys. Chem. B* 106, 3724-3734.
-D. Toptygin (2003) *J. Fluoresc.* 13, 201-219.

Date submitted: Editor Retained.

John M. Torkelson, Ph.D.

Dept. of Chemical Engineering,
Department of Materials Science and Engineering,
Northwestern University, Evanston, IL 60208-3120,
United States of America.
Tel: 847 491 7449
j-torkelson@northwestern.edu

Specialty Keywords: Polymers, Sensors, Glass Transition.

Fluorescence methods have been developed to address fundamental issues and applied problems in polymer science. These include the ability use ensemble and single-molecule fluorescence to quantify the effects of nanoscale confinement on the glass transition behavior, heterogeneous dynamics in polymers and nanocomposites, dye and polymer diffusion in polymers, conversion and block copolymer formation in reactive processing of polymers, and oxygen levels in pressure sensitive paints.

-J. C. Quirin and J. M. Torkelson (2003). Self-referencing sensor for monitoring conversion of nonisothermal polymerization and nanoscale mixing of resin components *Polymer* 44(2), 423-432.

Date submitted: 4th April 2006   **Jack T. Trevors, Ph.D.**

Department of Environmental Biology,
University of Guelph,
Guelph, Ontario,
Canada, N1G 2W1.
Tel: 519 824 4120 Ext: 53367   Fax: 519 837 0442
jtrevors@uoguelph.ca
www.oac.uoguelph.ca/env/bio/trevors/htm

Specialty Keywords: Polarization, Bacteriology.

Dr. Trevors current interests include: Membrane Fluidity, Reporter Genes. Cytoplasmic membrane polarization in bacteria. Fluorescent probes for detection of viable and non-viable bacterial cells.

-Pokorny, N., M. M. Hart, L. Storey, J. Boulter, H. Lee and J. T. Trevors. 2005. Hypobaric bacteriology: growth, cytoplasmic membrane polarization and total cellular fatty acids responses of *Escherichia coli* and *Bacillus subtilis*. Int. J. Astrobiol. 4:187-193.

-Denich, T. J., L. A. Beaudette, H. Lee and J. T. Trevors. 2005. Fluorescent methods to study DNA, RNA, proteins and cytoplasmic membrane polarization in the pentachlorophenol-mineralizing bacterium *Sphingomonas* sp. UG30 during nutrient starvation in water. J. Fluorescence. 15:143-151.

Date submitted: 30th June 2006   **Hira B.Tripathi, Ph.D.**

Photophysics Laboratory, Department of Physics,
Kumaon University, Naninital,
Uttranchal, PIN 236001,
India.
Fax: +91 594 223 5576
hiratripathi@yahoo.co.in

Specialty Keywords: Fluorescence, Time Domain Spectroscopy, Instrumentation.

Our research interest includes: fluorescence spectroscopy of some charge transfer complexes, excited state salvation dynamics, edge excitation red shift in emission, excited state proton transfer, energy transfer phenomena and their applications as optical sensors, lasing materials. We wish to undertake work on a time domain fluorescence microscopy. Edge excitation red shift and energy migration in quinine bisulphate dictation

-H. Mishra, Debi Pant, T.C. Pant and H. B. Tripathi, J. Photoch. Photobio. A Chemistry 177(2006)197-204.

-An experimental and Theoretical Investigation of the Photophysics of 1-Hydroxy-2-naphthoic Acid. H.Mishra, S. Maheswary, H. B. Tripathi and N. Sathyamurthy, J. Physical Chemistry A 2005, 109, 2746-2754.

Date submitted: 13th September 2006

Mahalingam Umadevi, Ph.D.

Department of Physics,
Mother Teresa Women's University,
Kodaikanal-624102, Tamil Nadu
India.
Tel: 91 944 392 8671   Fax: 91 454 224 1122
ums10@yahoo.com

Specialty Keywords: SERS, Fluorescence, Quinone derivatives.
AIM 2004 = 7.8

Optical absorption and fluorescence emission spectroscopic techniques have been employed to investigate the solvent effect, excited state interactions and inclusion of quinone derivatives in calix[8]arene. Orientations of biologically important quinone derivative on silver nano sol have been elucidated from the surface enhanced Raman measurements. Recently the preferential solvation of quinone derivatives in binary solvent mixture has been studied using spectroscopic techniques.

-M.Umadevi, A.Suvitha, K.Latha, Beulah J.M.Rajkumar, and V.Ramakrishnan (2006). Spectral investigations of preferential solvation and solute-solvent interactions of 1,4-dimethylamino anthraquinone in CH2Cl2/C2H5OH mixtures Spectrochim. Acta (In Press)

Date submitted: 15th July 2005

Evgenia Vaganova, Ph.D.

The Inorganic and Analytical Chemistry Department,
The Hebrew University,
Givat Ram, 94904, Jerusalem,
Israel.
Tel: 972 2 658 4199   Fax: 972 2 658 5319
gv@cc.huji.ac.il

Specialty Keywords: Photophysics, Photochemistry, Material Science.
AIM 2005 = 7.0

The goal of our research is focused on the understanding of the electronic properties of cumulative systems as in excited as well in a ground state; collective electronic behavior of the system, where mutual interaction can change the properties of the individual molecule.

-Vaganova E., Yitzchaik, S., Sigalov M., Borst, J.W., Visser, A., Ovadia, H., Khodorkovsky V., Time-resolved emission upon two-photon excitation of bis-N-carbazolyl-distyrylbenzene: mapping of water molecule distribution in the mouse brain, (2005) New Journal of Chemistry, in press, (advanced paper from 7/07/05).
-Vaganova, E., Rozenberg, M., Yitzchaik, S., Multicolor Emission in Poly(4-vinyl-pyridine) Gel, (2000) Chemistry of Materials, 12, 261-263.

Date submitted: 26th October 2006   **Jan Valenta, Ph.D.**

Department of Chemical Physics & Optics,
Faculty of Mathematics & Physics, Charles University,
Ke Karlovu 3, CZ-121 16 Prague 2,
Czech Republic.
Fax: +420 22 191 1249
jan.valenta@mff.cuni.cz
quantum.karlov.mff.cuni.cz/valenta

Specialty Keywords: Micro-spectroscopy, Nanocrystals, Photonic Structures, Scientific Photography.

Optical spectroscopy of individual low-dimensional semiconductor structures (nanocrystals - quantum dots) and biological complexes. Linear and non-linear optical properties of solids (pump-and-probe techniques, spectral hole-burning and filling, optical gain). Artificial and natural photonic structures. Scientific photography - optical imaging and processing.

-J. Valenta, J. Linnros, R. Juhasz, J.-L. Rehspringer et al.: Photonic band-gap effects on photoluminescence of silicon nanocrystals embedded in artificial opals, J. Appl. Phys. 93 (2003), 4471.

-I. Sychugov, R. Juhasz, J. Valenta, J. Linnros: Narrow luminescence linewidth of a silicon quantum dot, Phys. Rev. Lett. 94 (2005) 087405.

Date submitted: 28th April 2005   **Bernard Valeur, Ph.D.**

Conservatoire National des Arts et Métiers,
292 rue Saint-Martin,
75141 Paris Cedex 03,
France.
Tel: +33 (0)14 027 2622   Fax: +33 (0)14 027 2362
valeur@cnam.fr

Specialty Keywords: Fluorescent Molecular Sensors, Excitation Energy Transfer, Multichromophoric Systems.

Current interests: Supramolecular photophysics and photochemistry. Multichromophoric systems (e.g. antenna effect and energy hopping in multichromophoric cyclodextrins, calixarenes, porphyrin assemblies). Design of fluorescent molecular sensors for ion recognition (e.g. calixarene-based fluorescent sensors for the detection of alkali, alkaline-earth and heavy metal ions). Photoinduced motions of cations (e.g. photoejection from complexes).

-B. Valeur (2002). Molecular Fluorescence. Principles and Applications. Wiley-VCH, Weinheim.

-R. Métivier, I. Leray, B. Valeur (2004). Lead and mercury sensing by calixarene-based fluoroionophores bearing two or four dansyl fluorophores, *Chem.Eur. J.* 10, 4480-4490.

van Geest, L. K.
Van Sark, W.G.J.H.M.

Date submitted: 4th October 2006

Lambertus K. van Geest, M.Sc.

Lambert Instruments,
Turfweg 4,
9313 TH Leutingewolde,
The Netherlands.
Tel: +31 50 501 8461   Fax: +31 50 501 0034
lkvgeest@lambert-instruments.com
www.lambert-instruments.com

Specialty Keywords: FLIM, Image Intensifiers, ICCD Cameras, High Speed Cameras.

Development of imaging and detection systems for fluorescence microscopy often making use of image intensifiers which are fully digitally controlled and can be gated or modulated. Ongoing research that is aimed at the improvement of the instrument for Fluorescence Lifetime Imaging Microscopy (FLIM) and the application of LED's as modulated light source in such a system. New applications are developed in collaboration with universities and research institutes.

-L.K. van Geest, K.W.J. Stoop (2003). FLIM on a wide field fluorescence microscope, Published in "Letters in Peptide Science" Issue: Volume 10, Numbers 5-6 2003, Pages: 501 – 510 .

-K.W.J. Stoop, K. Jalink, S.J.G. de Jong, L.K. van Geest (2004) Measuring FRET in living cells with FLIM Proceedings of 8th Chinese International Peptide Symposium in press.

Date submitted: 12th October 2006

Wilfried G.J.H.M. Van Sark, Ph.D.

Department Science, Technology & Society,
Copernicus Institute, Utrecht University,
Heidelberglaan 2, 3584 CS Utrecht,
The Netherlands.
Tel: +31 30 253 7611   Fax: +31 30 253 7601
w.g.j.h.m.vansark@chem.uu.nl
www.chem.uu.nl/nws

Specialty Keywords: Quantum Dots, Solar Cells.

Having obtained research experience on solar cells materials development and nanocrystals I now combine these to develop new concepts of solar cells in which full spectral use is the challenge. Spectral converters, such as quantum dots, and down and up converter materials are employed and their performance is evaluated.

-W.G.J.H.M. van Sark *et al.* (2002). Blueing, bleaching, and blinking of single CdSe/ZnS quantum dots *ChemPhysChem* 3 871-879.

-W.G.J.H.M. van Sark (2005). Enhancement of solar cell performance by employing planar spectral converters, *Appl. Phys. Lett.* 87 151117.

M. J. vandeVen.
D. Vaudry.

Date submitted: Editor Retained.

Martin J. vandeVen, Ph.D.

Department MBW, Biomedical Research Institute (BIOMED) /
Institute of Materials Research (IMO),
Limburg University Center (LUC), Bldg D / Trans National,
University Limburg (tUL), University Campus,
Diepenbeek, B-3590, Belgium.
Tel: 0032 (0)11 268558 / 8816   Fax: 0032 (0)11 268599 / 8899
martin.vandeven@luc.ac.be
www.luc.ac.be/engels/ & www.tul.edu
Specialty Keywords: Spectrofluorimetry, Microscopy, Image Analysis.

Collaborative research centers on (1) fluorescence imaging microscopy of cellular interactions in autoimmune diseases Multiple Sclerosis (MS) and Rheumatoid Arthritis (RA) (2) polymer fluorescence characterization for biosensors (3) Chlorophyll and GFP fluorescence imaging related to leaf and fruit physiology (4) application of neural networks in image analysis (5) development of laser-based time- and frequency domain fluorescence methodologies at the LUC Biomed fluorescence Center.

-Using fluorescence images in classification of apples. Codrea, M.; Tyystjärvi, E.; vandeVen, M.; Valcke, R. and Nevalainen, O.; IASTED-VIIP Benalmadena, Malaga, Spain Sept. 9-12 2002.

Date submitted: 7th April 2006

David Vaudry, Ph.D.

European Institute for Peptide Research (IFRMP 23),
Laboratory of Cellular and Molecular Neuroendocrinology,
INSERM U413, UA CNRS,
University of Rouen 76821 Mont-Saint-Aignan Cedex, France.
Tel: 33 23 514 6760   Fax: 33 23 514 6946
David.vaudry@univ-rouen.fr
www.univ-rouen.fr/inserm-u413/microscopie.htm

Specialty Keywords: Cf - Microscopy, Time-laps Microscopy, Microarray, Proteomic, Q-RT-PCR & Microplate Reader.

We are conducting a functional evaluation of some genes and proteins that are regulates during cell death and differentiation (http://proger-cdd.crihan.fr).

-D. Vaudry et al. (2002) Pituitary adenylate cyclase-activating polypeptide protects rat cerebellar granule neurons against ethanol-induced apoptotic cell death. *Proc. Natl. Acad. Sci. USA* 99: 6398-6403.
-D. Vaudry et al. (2003) Regulators of cerebellar granule cell development act through specific signaling pathways. Science 300:1532-1534.

Date submitted: 2nd August 2004   **Jose Luis Vazquez-Ibar, Ph.D.**

Department of Molecular Biophysics and Biochemistry,
Yale University,
260 Whitney Av, BASS 433,
New Haven, 06250, USA.
Tel: 203 436 4894   Fax: 203 432 6946
vazquez@mail.csb.yale.edu

Specialty Keywords: Membrane Proteins, Luminescence, Tryptophan Oxidation.

My research is focused on the study of the structure and dynamics of membrane proteins by combining protein engineering and luminescence techniques. We have probed a stacking interaction between the substrate and a tryptophan residue in the lactose permease of *E. coli* by exploiting the luminescence properties of the indole ring. Also, we recently have demonstrated that H-bond interactions of a tryptophan residue can be studied by oxidizing the aromatic ring with n-bromosuccinimide.

-Vazquez-Ibar, J.L., Guan, L., Svrakic, M. & Kaback, H.R. (2003). Exploiting luminescence spectroscopy to elucidate the interaction between sugar and a tryptophan residue in the lactose permease of Escherichia coli. *Proc Natl Acad Sci U S A*. 100, 12706-11.

Date submitted: 6th April 2006   **Nikolai L. Vekshin, Ph.D.**

Russian Academy of Science,
Institute of Cell Biophysics,
Institutskaya Street -3, Pushchino,
Moscow region, 142290, Russia.
Tel: 095 923 7467 Ext: 292
nvekshin@rambler.ru
photonics.narod.ru

Specialty Keywords: Biophysics, Spectroscopy, Biopolymers.

N.L. Vekshin is a specialist in biological spectroscopy and photobiophysics, including: hypochromism of biopolymers, protein fluorescence, mechanisms of exciplex formation, resonance energy transfer, energy migration in DNA, fluorescent probes, fluorescence screening of drugs, photomodulation of enzymes and mitochondria, etc. He has many papers in journals, 4 books in the field of photonics of biostructures and 2 patents on multi-pass fluorescence cuvettes.
-Vekshin N.L. Photonics of Biopolymers (Springer, Berlin, 2002).

-Vekshin N.L. Energy Transfer in Macromolecules (SPIE, Bellingham, 1997).

Date submitted: Editor Retained.

Rance A. Velapoldi, Ph.D.

Nygaardskogen 28,
N-3408 Tranby,
Norway,

Tel: 0473 285 3445
velapoldi@netcom.no

Specialty Keywords: Fluorescence Standards, Corrected Spectra, Quantum Yields.

In late 60's and 70's, performed extensive research on organic species in solution and inorganic ion-doped glasses for use as macro- and micro-luminescence standards in addition to some analytical applications of fluorescence at the National Bureau of Standards, Washington, DC. (now NIST). Retired from NIST in 1999 but continuing research on standards and luminescence at the Pharmacy Institute, University of Oslo, Blindern, Norway.

-R.A. Velapoldi and K.D. Mielenz, NBS Special Publication 260-64, US Department of Commerce, Washington, DC. 1980.

-R.A. Velapoldi and M.S. Epstein, Luminescence Standards for Macro- and Microspectrofluorimetry; in "Luminescence Applications" M.C. Goldberg, Ed. ACS Symposium Series, 383, pp 98-126 (1989), Washington, DC.

Date submitted: Editor Retained.

Nel H. Velthorst, Ph.D.

Analytical Chemistry & Applied Spectroscopy, Laser Centre,
Vrije Universiteit Amsterdam,
de Boelelaan 1083, Amsterdam,
1081 HV, The Netherlands.
Tel: +31 20 444 7541   Fax: +31 20 444 7543
velthrst@chem.vu.nl
www.chem.vu.nl/acas/

Specialty Keywords: Laser Induced and High-resolution Molecular Fluorescence.

The research has been directed on the potential of laser-induced fluorescence detection coupled to CE and LC and on the development and application of Shpol'skii Spectroscopy and Fluorescence Line Narrowing Spectroscopy for identification in analytical and environmental analysis, in particular applied on polycyclic aromatic hydrocarbons and their metabolites.

-O.F.A. Larsen, I.S. Kozin, A.M. Rijs, G.J. Stroomberg, J.A. de Knecht, N.H. Velthorst and C. Gooijer: Direct identification of pyrene metabolites in organs of the isopod *Porcellio scaber* by Fluorescence Line Narrowing Spectroscopy. Anal. Chem. 70, 1182-1185 (1998).

Date submitted: Editor Retained.

Mariano Venanzi, Ph.D.

Department of Chemical Sciences and Technologies,
University of Roma Tor Vergata,
Via della ricerca scientifica, 00133, Roma,
Italy.
Tel: +39 067 259 4468   Fax: +39 067 259 4328
venanzi@uniroma2.it

Specialty Keywords: Biospectroscopy, Energy / Electron Transfer, Peptide Structure.

My research actvity focusses on the application of fluorescence spectroscopy and other photophysical techniques in the study of energy/electron flow in peptides and molecules of biological interest.   Current research projects: Structure of peptide foldamers; design and characterization of peptide-based molecular devices for memories, switches and energy conversion; energy/electron transfer in porphyrin dimers and dendrimers; photocatalysis in micelles and organized environments.

-(2002) Structural features and conformational equilibria of 310-helical peptides in solution by spectroscopic and molecular mechanics studies Biopolymers(Biospectroscopy) 67, 247-250.

-(2002) Effects of helical distortions on the optical properties of amide NH infrared absorption in short peptides in solution. J. Phys. Chem. B 106, 5733-5738.

Date submitted: Editor Retained.

Jo Vercammen, Ph.D.

Biochemistry, K.U. Leuven,
Celestijnenlaan,
Heverlee, 3001,
Belgium.
Tel: 00 321 632 7132   Fax: 00 321 632 7982
Jo.Vercammen@fys.kuleuven.ac.be

Specialty Keywords: HIV-1 integrase, Fluorescence Correlation Spectroscopy, Fluorescence Fluctuation Analysis.

The Laboratory of Biomolecular Dynamics is equipped with an FCS instrument and within this project this technique will be developed for the study of the enzyme integrase. HIV-1 integrase is a lentiviral protein and is regarded as one of the potential candidates for developing antiviral drugs, next to reverse transcriptase and protease. The study of the mechanism of the integrase reaction may also contribute to the further development of gene therapy using lentiviral vectors. The enzymatic activities of HIV-1 integrase will be studied as well as the multimerisation.

Date submitted: Editor Retained.

Antonie J. W. G. Visser, Ph.D.

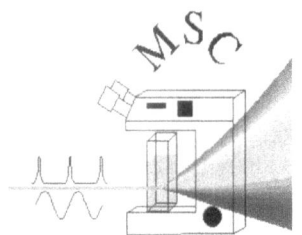

MicroSpectroscopy Centre, Wageningen University,
Dreijenlaan 3, 6703 HA Wageningen, The Netherlands.
And Department of Structural Biology, Vrije Universiteit,
De Boelelaan 1087, 1081 HV, Amsterdam, The Netherlands.
Tel: +31 31 748 2862   Fax: +31 31 748 4801
Ton.Visser@laser.bc.wau.nl
www.mscwu.nl

Specialty Keywords: Flavoproteins, Fluorescence Fluctuations,
Time-resolved Fluorescence.

The mission of our MicroSpectroscopy Centre is to strengthen the Dutch infrastructure in optical microspectroscopy, in particular fluorescence. We offer universities, research institutes and industrial companies microspectroscopic state-of-the-art facilities in which biomolecular interactions can be studied such as those among proteins, carbohydrates, lipids, nucleic acids, metabolites, either in isolated form or within cells. Current research is focused on: signal transduction in plants; characterization of plant pathogen resistance genes; gene display technology with high throughput screening; redox biochemistry in complex media and characterization of mesoscopic systems in food sciences.

Structural dynamics of green fluorescent protein alone and fused with a single chain Fv protein.
-M.A. Hink, R.A. Griep, J.W. Borst, A. van Hoek, M.H.M. Eppink, A. Schots and A.J.W.G. Visser (2000) J. Biol. Chem. 275, 17556-17560.

Date submitted: Editor Retained.

Radka S. Vladkova, Ph.D.

Section Model Membranes, Institute of Biophysics,
Bulgarian Academy of Sciences,
Acad. G. Bonchev Str., Bl. 21, Sofia 1113,
Bulgaria.
Tel: +359 2 979 3694   Fax: +359 2 971 2493
vladkova@obzor.bio21.bas.bg
www.bio21.bas.bg/ibf

Specialty Keywords: Chlorophylls, Membranes, Fluorescent
Probes.

Intermolecular interactions, organization and dynamics of both the fluorescing molecules (e.g. Chlorophylls, 1,8-ANS) and the medium where they are imbedded (solvents, mixtures, low-temperature matrices, membrane lipid-water structures, and photosynthetic membranes) by using the full arsenal of fluorescence characteristics estimated from steady-state and time-resolved emission spectroscopy, as well as those from hole-burning and site-selection spectroscopy.

-R. Vladkova (2000). Chlorophyll *a* self-assembly in polar solvent-water mixtures. *Photochem. Photobiol.*, 71(1), 71-83.

-R. Vladkova, K. Teuchner, D. Leupold, R. Koynova and B. Tenchov (2000). Detection of the metastable rippled gel phase in hydrated phosphatidylcholine by fluorescence spectroscopy. *Biophys. Chem.*, 84(2), 159-166.

Date submitted: Editor Retained.

Peter Vöhringer, Ph.D.

Max-Planck-Institute for Biophysical Chemistry,
Biomolecular and Chemical Dynamics Group,
Am Fassberg 11, Göttingen, 37077,
Germany.
Tel: +49 (551) 201 1333   Fax: +49 (551) 201 1341
pvoehri@gwdg.de
www.mpibpc.gwdg.de/abteilungen/072

Specialty Keywords: Femtosecond Spectroscopy, Condensed Matter.

Current interests: Dynamics of structural relaxations in biological environments. Ultrafast primary events involved in bioluminescence. Proton, electron, and energy transfer in condensed phase systems. Coherence in liquid phase chemical reactions. Molecular dynamics of liquids.

-K. Winkler, J. Lindner, and P. Vöhringer (2002) Low-frequency depolarized Raman-spectral density from femtosecond optical Kerr-effect experiments: Lineshape analysis of restricted translational modes, *Phys. Chem. Chem. Phys.* 4, 2144-2155.
-K. Winkler, J. Lindner, V. Subramaniam, T.M. Jovin, and P. Vöhringer (2002) Ultrafast dynamics in the excited state of green fluorescent protein (wt) studied by frequency-resolved femtosecond pump-probe spectroscopy, *Phys. Chem. Chem. Phys.* 4, 1072-1081 (2002).

Date submitted: 5th April 2006

Anna von Mikecz, Ph.D.

Inst. umweltmed. Forschung, Heinrich-Heine-University,
Auf'm Hennekamp 50,
40225 Duesseldorf,
Germany.
Tel: +49 211 338 9358
mikecz@uni-duesseldorf.de
www.tec-source.de/mikecz

Specialty Keywords: Cell Nucleus, Confocal Microscopy, Proteasomal Proteolysis, Subnuclear Pathology of Disease.

The mammalian cell nucleus is composed of dynamic subnuclear compartments that form in response to gene expression ($\rightarrow$ form follows function). Disruption of nuclear function by xenobiotics such as heavy metals and nanoparticles results in altered proteasomal degradation and protein aggregation within the nucleus. These subnuclear pathologies occur in cellular senescence, neurodegenerative diseases and systemic autoimmune disorders, and can be visualized by confocal laser scanning microscopy. The nuclear ubiquitin-proteasome system plays a dual role: regulation of function and disease pathology in the cell nucleus.

-von Mikecz, A. The nuclear ubiquitin-proteasome system (nUPS). *J Cell Science* (in press).

Date submitted: 28th April 2006

Alan S. Waggoner, Ph.D.

Carnegie Mellon University,
Molecular Biosensor and Imaging Center,
4400 Fifth Avenue,
Pittsburgh, Pennsylvania, 15213-2683, USA.
Tel: 412 268 3461 Fax: 412 268 6571
waggoner@andrew.cmu.edu
www.cmu.edu/bio/faculty/waggoner.html

Specialty Keywords: Fluorescence, Probes, Microscope Imaging.

Development and application of fluorescence technologies in basic biological research, biotechnology, and medical diagnostics. This work includes new multicolor fluorescent labeling reagents, multi-parameter fluorescent antibodies, DNA probes, physiological indicators, molecular biosensors and associated fluorescence imaging systems.

-Waggoner AS. Fluorescent Labels for Proteomics and Genomics Current Opinion in Chemical Biology. 10:62-66 (2006).

-Waggoner AS. Fluorescent Labels for Proteomics and Genomics Current Opinion in Chemical Biology. 10:62-66 (2006).

Date submitted: 28th June 2005

Michael Wahl, Ph.D.

PicoQuant GmbH.,
Rudower Chaussee 29,
12489 Berlin,
Germany.
Tel: +49 306 392 6562 Fax: +49 306 392 6561
wahl@pq.fta-berlin.de
www.picoquant.com

Specialty Keywords: TCPC, Time-resolved Fluorescence, Fluorescence Correlation, Single Molecules.

M.W. is working as senior scientist and project leader at PicoQuant GmbH. His research focuses on instrumentation and data analysis software for time-correlated single photon counting. These instruments are applied in ultra-sensitive analysis down to the single molecule level. Recent projects include novel data acquisition systems for time-resolved fluorescence microscopy and advanced data analysis algorithms for fluorescence correlation spectroscopy and fluorescence lifetime imaging, as well as work towards a new high resolution TCSPC system and picosecond correlator.

-A. Benda, M. Hof, M. Wahl, M. Patting, R. Erdmann, and P. Kapusta (2005). TCSPC upgrade of a confocal FCS microscope *Rev. Sci. Instrum.* 76, 033106.

Date submitted: 25th June 2005

Lai-Hao Wang, Ph.D.

Applied Chemistry,
Chia Nan University of Pharmacy and Science,
Tainan,
Taiwan R.O.C.
Tel: 886 7 801 8515   Fax: 886 6 266 7319
c8010111@mail.chna.edu.tw

Specialty Keywords: Fluorescence.

In our research, for determination of metabolites in human biological fluids the time course of percutaneous absorption and orally drug, have been investigated using fluorometric or electrochemical detection with high-performance liquid chromatography and our own construction flow systems.

-Determination of 5-methoxypsoralen in human serum, Lai-Hao Wang and Mey Tso,*J. Pharm.& Biomed. Anal.*, 2002, 30,593.

-Determination of urinary metabolites of coumarin in human urine by HPLC, Lai-Hao Wang, Chia-Ling Lien, *J. Liquid Chromarography & Related Technologies*, 2004,27, 1.

Date submitted: 30th May 2006

William W. Ward, Ph.D.

Biochemistry & Microbiology,
Rutgers University, Cook College,
76 Lipman Dr., New Brunswick,
New Jersey, 08901 USA.
Tel: 732 932 9562 ext 216   Fax: 732 932 3633
crebb@rci.rutgers.edu
www.rci.rutgers.edu/~meton/protein.html

Specialty Keywords: GFP, Bioluminescence, Proteins.
Professor Ward specializes in physical and chemical properties of green-fluorescent protein (GFP). He teaches a GFP-based short course in protein purification at his center (CREBB) and has organized GFP symposia in 1997, 1999, and 2004. He also teaches "Fluorescence:Basic Principles and Applications in Drug Discovery" for IBC. He has published more than 100 refereed papers, book chapters, and abstracts and has one issued patent and one pending on HTS of cell-based GFP.
-Ward, W.W., Biochemical and Physical Properties of Green Fluorescent Protein (1998) in Green Fluorescent protein, M. Chalfie and S.Kain, eds. Wiley-Liss pp 45-75.
-H.A.Richards, C-T.Han, R.G.Hopkins, M.L.Failla,W.W.Ward, and C.N.Stewart(2003), Safety Assessment of Recombinant Green Fluorescent Protein Orally Administered to Weaned Rats, J.Nutr.,133:1909-1912.

Date submitted: Editor Retained.

Peter Wardman, Ph.D., D.Sc.

Cancer Research UK *Free Radicals* Research Group,
Gray Cancer Institute, PO Box 100, Mount Vernon Hospital,
Northwood, Middx HA6 2JR,
U.K.
Tel: +44 (0)192 382 8611   Fax: +44 (0)192 383 5210
wardman@gci.ac.uk
www.gci.ac.uk

Specialty Keywords: Free radicals, Oxidative stress, Radiation chemistry.

My interests focus on the roles of free radicals in cancer biology, particularly the chemistry of cellular oxidative stress and the detection of free radicals or their products in biological systems. Radiation-produced free radicals are of special interest, as are the kinetics of radical reactions. Pulse radiolysis, stopped-flow rapid-mixing and EPR are to characterize reaction kinetics. The chemistry of fluorescent probes that are putative 'reporters' of oxidative and nitrosative stress is of current interest.

-Wardman, P., *et al.*, 2002, Pitfalls in the use of common luminescent probes for oxidative and nitrosative stress. *Journal of Fluorescence*, 12, 65-68.

-Ford, E., *et al.*, 2002, Kinetics of the reactions of nitrogen dioxide with glutathione, cysteine, and uric acid at physiological pH. *Free Radical Biology & Medicine*, 32, 1314-1323.

Date submitted: 8th October 2006

Watt W. Webb, Sc.D.

Applied & Engineering Physics, Cornell University,
223 Clark Hall,
Ithaca, NY 14853,
USA.
Tel: 607 255 3331   Fax: 607 255 7658
www2@cornell.edu
www.aep.cornell.edu/drbio

Specialty Keywords: Biophysics, Biomedical, Optics.
AIM 2005 = 26.2

The aim of our research is to understand, at the molecular level, the dynamics of basic biophysical processes. The continual challenge is to detect the exquisite subtlety of biomolecular signals and to broaden the paradigms of physical science to encompass biological complexity. The creation of new physical instrumentations addresses this challenge. We study the dynamics of biophysical processes in living cells using modern physical optics such as fluorescence correlation spectroscopy and nonlinear laser scanning microscopy.

-Magde, D., Webb, W. W. & Elson, E. Thermodynamic Fluctuations in a Reacting System - Measurement by Fluorescence Correlation Spectroscopy. *Physical Rev Lett* 29, 705-708 (1972).

-Denk, W., Strickler, J. H. & Webb, W. W. Two-Photon Laser Scanning Fluorescence Microscopy. *Science* 248, 73-76 (1990).

Date submitted: 16th October 2006

Wilfried Weigel, Ph.D.

Institute of Chemistry,
Humboldt University Berlin,
Brook-Taylor-Str. 2, 12489 Berlin,
Germany.
Tel: 49 (0) 302 093 5583   Fax: 49 (0) 302 093 5574
weigel@chemie.hu-berlin.de
www.chemie.hu-berlin.de/wr/group/willi.html

Specialty Keywords: Fluorescence Sensing, Time-resolved Fluorescence, Functional Surface Coating.

Current Research: Development of fluorescent probes, photophysical relaxation mechanisms, application of probes in dendritic structures, fluorescence behavior of immobilized dyes, characterization of surface functionalities with fluorescence methods.

-W. Weigel et al. (2003). Dual Fluorescence of Phenyl and Biphenyl Substituted Pyrene Derivatives: *J. Phys. Chem. A* 107, 5941.

-W. Weigel et al. (2005). Phenylene Alkylene Dendrons with Site-Specific Incorporated Fluorescence Pyrene Probes: *J. Org. Chem.* 70, 6583.

Date submitted: 27th October 2006

Klaus D. Weißhart, Ph.D.

Carl Zeiss MicroImaging GmbH.,
Advanced Imaging Microscopy,
Carl-Zeiss-Promenade 10,
07745 Jena, Germany.
Tel: +49 364 164 2268   Fax: +49 364 164 3144
weisshart@zeiss.de
www.zeiss.de

Specialty Keywords: Fluorescence Correlation Spectroscopy (FCS), Single Molecule Detection (SMD).

Responsibilities: Product management for the ConfoCor product line of Carl Zeiss used for fluorescence correlation spectroscopy, single molecule detection and highly sensitive imaging. Interests: Development of advanced instrumentation to improve these technologies.

-Carl Zeiss MicroImaing GmbH., provides customers in life sciences with a versatile program of light optical microscopes and systems for image processing and documentation, for laser scanning microscopy and fluorescence correlation spectroscopy.

Date submitted: Editor Retained.

Gunnar Westman, Ph.D.

Department of Chemistry and Bioscience,
Chalmers University of Technology,
S-412 96 Gothenburg,
Sweden.
Tel: 46 31 772 3072   Fax: 46 31 772 3657
westman@oc.chalmers.se
www.oc.chalmers.se

Specialty Keywords: Synthesis, Cyanine Dyes, Benzophenoxazine.

Current interests: Design and synthesis of new fluorescent molecules for the detection and studies of biological systems. Currently we design fluorescent probes that bind in the minor groove of nucleic acids. We also develop fluorescent dyes that show specific staining of cells.

-Svanvik, N., Westman, G., Wang, D. and Kubista M. Anal. Biochem. 281, 26-35 (2000).

-Isacsson J and Westman G Tet. Lett. 42, 3207-3210 (2001).

Date submitted: Editor Retained.

Jerker Widengren, Ph.D., M.D.

Dept. Medical Biophysics, MBB,
Scheeles v. 2, Karolinska Institutet,
171 77 Stockholm,
Sweden.
Tel: +46 8 728 6815   Fax: +46 832 6505
jerker.widengren@mbb.ki.se

Specialty Keywords: FCS, Single Molecule Spectroscopy.

Current research: Development of techniques and applications of Fluorescence Correlation Spectroscopy (FCS) and single molecule Multi-parameter Fluorescence Detection (smMFD). Monitoring and characterization of transient photophysical states, conformations and conformational fluctuations of biomolecules. Detection, characterization and diagnostics of sparse amounts of biomolecules on cell surfaces and in body fluids.

-Widengren J, Schweinberger E, Berger S, and Seidel C: J. Phys. Chem., 105, 6851-6866, 2001.
-Widengren J, Mets, Ü: Conceptual basis of FCS and related techniques as tools in bioscience p. 69-119, in Single Molecule Detection in Solution, Eds. Zander, Enderlein, Keller, Wiley VCH 2002.

Date submitted: 30th June 2005

Léonard Widmer.

idQuantique SA,
Chemin de la Marbrerie 3,
1227 Carouge-Geneva,
Switzerland.
Tel: +41 22 301 8371   Fax: +41 22 301 8379
leonard.widmer@idquantique.com
www.idquantique.com

Specialty Keywords: TCSPC, Single Photon Detection, Fluorescence, Spectroscopy, FLIM, Transillumination Imaging.

Leonard Widmer is a Specialist of single photon detectors and their uses in chemical, biological and medical applications. Special interests in high resolution TCSPC experiments.

Date submitted: 12th October 2006

VARIAN

Gert J. Wilgenhof, Ing.

Varian B.V.,
Herculesweg 8, Middelburg,
Postbus 8005, 4330 EA,
The Netherlands.
Tel: +31 11 867 1500   Fax: +31 11 867 1502
Gert.Wilgenhof@Varianinc.com
www.varianinc.com

Specialty Keywords: Fluorometer, Spectrofluorometer, Applications.

Varian offers high quality products for measuring fluorescence in many applications. Especially the Cary Eclipse fluorometer offers every wavelength for analyzing fluorescence, phosphorescence and chemi-luminescence with excitation and emission scans as well as 2D / 3D plots. Temperature control, polarization, fiber optic and wellplate options are available. With the instrument knowledge Varian participates in research projects and helps with developing new applications. The Varian office in Middelburg is equipped with all the necessary tools to make your fluorescent application work.

Please contact: Gert Wilgenhof – Cary Eclipse specialist The Netherlands.

Date submitted: 25th February 2005

Gerald M. Wilson, Ph.D.

Department of Biochemistry and Molecular Biology,
University of Maryland School of Medicine,
108 N. Greene Street, Baltimore, Maryland 21201,
USA.
Tel: 410 706 8904   Fax: 410 706 8297
gwils001@umaryland.edu

Specialty Keywords: RNA biology, RNA folding, FRET.

My principal research foci concern trans-acting factors contributing to the regulation of messenger RNA turnover and the roles of RNA conformational heterogeneity in modulating association and function of these factors. To these ends, we employ fluorescence anisotropy and resonance energy transfer to evaluate RNA-protein binding and RNA folding events under solution conditions.

-Wilson, G.M., Lu, J., Sutphen, K., Suarez, Y., Sinha, S., Brewer, B., Villanueva-Feliciano, E.C., Ysla, R.M., Charles, S., and Brewer, G. (2003) *J. Biol. Chem.*, 278, 33039-33048.
-Brewer, B.Y., Malicka, J., Blackshear, P.J., and Wilson, G.M. (2004) *J. Biol. Chem.*, 279, 27870-27877.

Date submitted: 20th June 2004

Otto S. Wolfbeis, Ph.D.

Institute of Analytical Chemistry,
Chemo- and Biosensors,
93040 Regensburg,
Germany.
Tel: +49 941 943 4066   Fax: +49 941 943 4064
otto@wolfbeis.de
www.wolfbeis.de

Specialty Keywords: FLIM, Fluorescent Probes, Fluorescent (bio)sensors, New Materials for Sensors.
AIM 2003 = 44.6

Main research interests: (fiber optic) chemical sensing and biosensing; novel schemes in analytical fluorescence spectroscopy incl. dual lifetime referencing (DLR); design of advanced materials for use in (bio)chemical sensing; clinical sensing; conducting organic polymers for sensors; fluorescent probes derived from ruthenium and europium; protein and DNA labels; fluorescent beads; biosensors based on thin gold films and molecular imprints;
-Fiber Optic Chemical Sensors and Biosensors (2002-2003), O. S. Wolfbeis, *Anal. Chem.* 2004, *76*, 3269-3283 (biannual review).
-Advanced Luminescent Labels, Probes and Beads, and Their Application to Luminescence Bioassay and Imaging, O. S. Wolfbeis et al., *Springer Series in Fluorescence Spectroscopy,* vol. 2 (R. Kraayenhof, A. J. W. G. Visser, H. C. Gerritsen, eds.), Springer, Berlin, 2002; pp. 3-42.

Date submitted: 22nd May 2005

Danuta Wróbel, Ph.D.

Institute of Physics, Poznan University of Technology,
Nieszawska 13A, Poznan,
Poland, 60-965,
Poland.
Tel: +48 61 669 3179   Fax: +48 61 669 3201
wrobel@phys.put.poznan.pl
www.put.poznan.pl
Specialty Keywords: Molecular Physics, Molecular
Spectroscopy, Organic Dyes.
AIM 2003 = 5.6

The study of spectral properties of synthetic organic dyes and chlorophyll pigments in isotropic and anisotropic media to follow: Mechanisms of radiative and non-deactivation processes of porphyrins and phthalocyanines, of porphyrin-melanin systems, mechanisms of generation of the photovoltaic effects in photoelectrochemical cells based on synthetic organic dyes and biological materials, Langmuir-Blodgett layers, optical and IR studies, organic photovoltaics, photodynamic therapy.

-D.Wróbel, et al., Fluorescence and time-resolved delayed luminescence of porphyrins in organic solvents and polymer matrices, J.Fluorescence, 8 (1998) 191-198.
-A.Boguta, D.Wróbel, Fluorescein and phenolophthalein-Correlation of fluorescence and photovoltaic properties, J. Fluorescence, 11 (2001), 131-139.

Date submitted: 13th October 2006

Tao T. Xu, Ph.D.

Academy of Metrology and Quality Inspection,
Longzhu Road, Shenzhen,
Nanshan, 518055,
China.
Tel: 86 07 552 694 1524
xutao780606@126.com

Specialty Keywords: Biomedical optics, Fluorescence spectrum,
Optical measurement.

Dr Xu's current interests are, medical lasers, tissue fluorescence, tissue optics and optical measurement etc. He has worked and is working on the oxygen variance in the photodynamic therapy of tumor, the cancer therapy effect by different laser fluence rate in PDT, the difference fluorescence spectrum of normal tissue and tumor and the diffuse reflectance of the silicon photodiode etc

-Yuezhi Li, Mingzhao Li, Tao Xu (2006) Quantitative Time-Resolved Fluorescence Spectra of the Cortical Sarcoma and the Adjacent Normal Tissue Determined with an in Vivo Experimental Method and Theoretical Model. Applied spectroscopy 60: 808-812.

Date submitted: Editor Retained.

Li Yao-Qun, Ph.D.

Department of Chemistry,
Xiamen University,
Xiamen 361005,
China.
Tel / Fax: +86 592 218 5875
yqlig@xmu.edu.cn

Specialty Keywords: Fluorescence, Multi-component Analysis.
The research fields include molecular fluorescence spectroscopy and its application in environmental and biological analysis, multi-component analysis, and surface analysis. Special interests have focused on the development, instrumentation and application of some fluorescence techniques, such as synchronous fluorescence spectroscopy, multi-dimensional fluorescence, derivative technique, reflection fluorescence and confocal microscopy.
-Derivative matrix isopotential synchronous fluorescence spectroscopy for the direct determination of 1-hydroxypyrene as a urinary biomarker of exposure to polycyclic aromatic hydrocarbon, with Wei Sui, Chun Wu, Li-Jun Yu, *Anal Sci.*, 2001, 17(*1*),167.
-Spectral fluctuation and heterogeneous distribution of porphine on the water surface, with Maxim. N. Slyadnev, Takanori Inoue, Akira Harata and Teiichiro Ogawa, *Langmuir*,1999, 15(*9*), 3035.

Date submitted: 19th October 2006

Sergiy M. Yarmoluk, Ph.D.

Inst. of Molecular Biology and Genetics of NAS of Ukraine,
Zabolotnogo Str. 150,
Kyiv, 03143,
Ukraine.
Tel / Fax: +38 044 522 2458
sergiy@yarmoluk.org.ua
www.yarmoluk.org.ua

Specialty Keywords: Organic Synthesis, Fluorescent Probes, Cyanines
The research interests of Prof. Yarmoluk are connected with fluorescent detection of biological molecules. In the Department of Combinatorial Chemistry guided by Prof. Yarmoluk the series of novel dyes promising for use as fluorescent probes for nucleic acids and proteins detection were created [1,2], and novel methods for biomolecules labeling with cyanine dyes were developed. Also mechanisms of interaction of the dyes with nucleic acids and photophysical properties of the dyes are studied.
-V.B. Kovalska, V.P. Tokar, M.Yu. Losytskyy, T. Deligeorgiev, A. Vassilev, N. Gadjev, K.-H. Drexhage, S.M. Yarmoluk. (2006) J. Biochem. Biophys. Methods, *68*, 155-165.
-T.Y. Ohulchanskyy, H.E. Pudavar, S.M. Yarmoluk, V.M. Yashchuk, E.J. Bergey, P.N. Prasad (2003) Photochemistry and Photobiology, *77*, 138-145.

Date submitted: 7th April 2006

Hartmut Yersin, Ph.D.

Institut für Physikalische Chemie,
Universität Regensburg,
D-93040 Regensburg,
Germany.
Tel: +049 941 943 4464
hartmut.yersin@chemie.uni-regensburg.de
www.uni-regensburg.de/~hartmut.yersin

Specialty Keywords: Phosphorescent Compounds, Triplet Emitters, Development of OLED Materials.

Emission properties of organo-transition-metal compounds are controlled by chemical means. In particular, chemically induced tuning of the $^3LC(\pi\pi^*)/^3MLCT(d\pi^*)$ character results in changes of transition energy, spectral distribution, emission decay time, quantum yield, etc. Detailed spectroscopic investigations ($1.2 \leq T \leq 300$ K; $0 \leq$ pressure ≤ 60 kbar; $0 \leq$ magnetic field ≤ 12 T; highly resolved and time resolved emission etc.) lead to a deeper understanding of material's properties. [1] Suitable materials are applied in electro-luminescent devices (OLEDs). [2, 3]
-H. Yersin, D. Donges (2001); Top. Curr. Chem. 214, 81-186.
-H. Yersin (2004); Top. Curr. Chem. 241, 1-26.
-H. Yersin; "Highly effizient OLEDs with phosphorescent materials", Wiley-VCH, Weinheim (2006).

Date submitted: 7th July 2004

Lu Yi, Ph.D.

Department of Chemistry,
University of Illinois at Urbana - Champaign,
600 S. Mathews Ave., Urbana, IL 61801,
U.S.A.
Tel: 217 333 2619   Fax: 217 333 2685
yi-lu@uiuc.edu
montypython.scs.uiuc.edu/

Specialty Keywords: FRET, Biosensor, DNA.

We are interested in the design of catalytic DNA-based fluorescent biosensors for a broad range of non-nucleic acid analytes such as metal ions [1]. The substrate and enzyme strands of catalytic DNA were labeled with fluorophore and quencher, respectively, resulting in a suppressed initial fluorescence. In the presence of target analyte, the substrate was cleaved by the enzyme, increased fluorescence was observed. We have also developed a multi-fluorophore FRET method to label several arms of biomolecules simultaneously and study their folding [2].
- J. Li and Y. Lu (2000). A Highly Sensitive and Selective Catalytic DNA Biosensor for Lead Ions *J. Am. Chem. Soc.* 12210466-10467.
- J. Liu and Y. Lu (2002). FRET Study of a Trifluorophore-Labeled DNAzyme *J. Am. Chem. Soc.* 124(51), 15208-15216.

Date submitted: 5th April 2006

Liming Ying, Ph.D.

Department of Chemistry,
University of Cambridge,
Lensfield Road, Cambridge, CB2 1EW,
United Kingdom.
Tel: +44 122 333 0107   Fax: +44 122 333 6362
ly206@cam.ac.uk
www.ch.cam.ac.uk/CUCL/staff/ly.html
Specialty Keywords: Single Molecule Biophysics,
Nanobiotechnology, Biophysical Chemistry.
AIM 2005 = 15.2

My research focuses on the emerging multidisciplinary field of single molecule biophysics. I am developing and applying ultra-sensitive fluorescence spectroscopy tools to address fundamental questions in chemistry and molecular biology that are not accessible by conventional ensemble measurements.

-L. M. Ying, A. Bruckbauer, D. J. Zhou, J. Gorelik, A. I. Shevchuk, M. Lab, Y. E. Korchev, D. Klenerman. (2005) The scanned nanopipette: a new tool for high resolution bioimaging and controlled deposition of biomolecules, *Phys. Chem. Chem. Phys.* 7, 2859-2866.

Date submitted: 2nd March 2004

Xianghua (Bruce) Yu, Ph.D.

Department of Chemistry,
University of Tennessee,
Knoxville, Tennessee 37996,
USA.
Tel: 865 974 3738   Fax: 865 974 3454
yu@novell.chem.utk.edu
web.utk.edu/~xyu1

Specialty Keywords: Organometallics, X-ray, Synthesis.

Materials containing both metals and silicon are important components of current very-large-scale-integration (VLSI) devices. These materials include M-Si-N ternary materials and metal silicides. Our research focuses on synthesis, structures and reactivity of groups 4 and 5 transition metal amide silyl complexes. Their kinetics and thermodynamics have also been studied. I have also served as a group single crystal X-ray crystallographer for three years and solved about fifty-five new organic and organometallic complexes.

-X. Yu, H. Cai, I. A. Guzei, Z. Xue. (2004). Unusual equilibria involving group 4 amides, silyl complexes, and silyl anions via ligand exchange reactions. *J. Am. Chem. Soc.* in press.
-X. Yu, F. Li, X. Ye, X.; Xin, Z. Xue. (2000). Synthesis of cerium(IV) oxide ultrafine particles by solid-state reactions. *J. Am. Ceram. Soc. 83*, 964.

Date submitted: 14th June 2005

Jingli Yuan, Ph.D.

Dept. of Analytical Chem., Dalian Inst. of Chemical Physics,
Chinese Academy of Sciences,
457 Zhongshan Road, Dalian 116023
P. R. China.
Tel: / Fax: +86 41 18 437 9660
jingliyuan@yahoo.com.cn
www.gsc.dicp.ac.cn/dsjs/fx1/yjl.htm
Specialty Keywords: Lanthanide, Fluorescence Probe,
Bioassay.
AIM 2004 = 17.4

Several kinds of lanthanide chelate-based fluorescent nanoparticles have been prepared and developed as new type of fluorescence probes for biolabeling and time-resolved fluorescence bioassay. As fluorescence probes, the newly developed lanthanide nanoparticles have the advantages of smaller size (< 50 nm), high hydrophilicity, biocompatibility, photo-stability and fluorescence quantum yield (10-50%), easy to be bound to biomolecules and used for highly sensitive time-resolved fluorescence bioassays.

-Z. Ye, M. Tan, G. Wang and J. Yuan (2004) *Anal. Chem.* 76, 513-518.
-M. Tan, G. Wang, X. Hai, Z. Ye and J. Yuan (2004) *J. Mater. Chem.* 14, 2896-2901.

Date submitted: 19th July 2005

Il Yun, Ph.D.

Department of Dental Pharmacology and Biophysics,
College of Dentistry and Research Inst. for Oral Biotechnology,
Pusan University, 1-10 Ami-dong, Seo-gu, Busan,
602-739, Republic of Korea.
Tel: 81 51 240 7813   Fax: 82 51 254 0576
iyun@pusan.ac.kr

Specialty Keywords: Membrane-drug Interactions, Fluorescence
Probe Technique, Membrane Fluidity.
AIM 2003 = 13.4

I am studying biophysical interactions between drugs and biomembranes or liposomes with the fluorescent analysis using fluorescent probes. Utilizing the fluorescent analysis, I have studied biophysical interactions between drugs such as local anesthetics, n-alkanols, barbiturates, chlorhexidine, dopamine, chlorpromazine, and catechin and synaptosomal plasma membrane vesicles (SPMV) which were isolated from bovine cerebral cortex and liposomes (model membranes) formed by phospholipids.

-I. Yun, E.S. Cho, H. O. Jang, U. K. Kim, C. H. Choi, I. K. Chung, I. S. Kim, W. G. Wood (2002) Amphiphilic effects of local anesthetics on rotational mobility in neuronal and model membranes. *Biochim. Biophys. Acta* 1564, 123-132.

Date submitted: 6th April 2006

Urszula Zabarylo, (Ph.D. Candidate)

Charité – Universitätsmedizin Berlin,
Campus Benjamin Franklin,
Biomedizinische Technik und Physik,
Fabeckstr. 60-62, 14195 Berlin, Germany,
Tel.: 49 308 449 2371   Fax: 49 308 445 4377
urszula.zabarylo@charite.de
www.charite.de

Specialty Keywords: Optical Molecular Imaging, Rheumatoid Arthritis, transillumination, Image processing and analysis.

Current Research Interests: My research is a development of novel laser-based imaging technique for multispectral diagnosis of early rheumatoid arthritis of the finger joint tissue. Special research interests are evaluation an optimal preprocessing method for an optical classification of pathological finger joints. Special interest is a development of scattering light images preprocessing algorithms like a deconvolution, which would be used to improve sensitivity and specifity of experimental diagnosis in medicine using conventional image processing methods.
-Minet O, Zabarylo U, Beuthan J: Deconvolution of laser based images for monitoring rheumatoid arthritis. Las. Phys. Lett. 2 (2005) 556-565.

Date submitted: 6th August 2005

Sergei Yu. Zaitsev, Ph.D.

Organic and Biological Chemistry,
Moscow State Acad. of Veterinary Medicine & Biotechnology,
Scryabin Str. 23, Moscow 109472,
Russia.
Tel: +70 95 377 9539   Fax: +70 95 377 4939
szaitsev@mail.ru

Specialty Keywords: Polymer Films, Fluorescent Crown-ether Dyes, Biomembranes.

Main research topics: Supramolecular chemistry, design and application of composite polymeric materials and sensors. Membrane mimetic chemistry (monolayers, Langmuir-Blodgett films, BLM, liposomes, micelles). Chemistry of lipids and membrane proteins, biomembranes. Synthesis and properties of surface-active monomers and polymers.
-Turshatov A. A., Bossi M., Möbius D., Hell S., Vedernikov A. I., Gromov S. P., Lobova N. A., Alfimov M. V., Zaitsev S. Yu., (2005), Photosensitive Properties Of A Novel Amphiphilic Dye, Thin Solid Films 476, 336-339.
-Zaitsev S. Yu., Sergeeva T. I., Baryshnikova E. A., Gromov S. P., Fedorova O. A., Alfimov M. V., Hacke S., Möbius D. (2002), Anion-Capped Benzodithia-18-Crown-6 Styryl Dye Monolayers , Colloids & Surfaces. A 198-200, 473-482.

Date submitted: 4th April 2006

Christoph C. Z. Zander, Ph.D.

AG Biochemistry, University of Siegen,
Adolf-Reichwein-Strasse 2, Siegen,
D-57068,
Germany.
Tel: +49 271 740 4478
zander@chemie.uni-siegen.de
www.uni-siegen.de

Specialty Keywords: Laser Cooling, Single Molecule Detection, Anti Stokes Fluorescence.

My group is working since 1991 in the field of fluorescence. The mayor topics of this works are laser cooling by anti Stokes fluorescence and single molecule detection (see ref. 1).

-Single Molecule-Detection in Solution: Methods and Applications, eds. Ch. Zander. J. Enderlein, R.A. Keller, Wiley-VCH Verlag Berlin GmbH, S. 247 – 272, Berlin 2002.

Date submitted: 12th October 2006

Jian Zhang, Ph.D.

Center for Fluorescence Spectroscopy,
Dept. of Biochemistry and Molecular Biology,
University of Maryland School of Medicine,
725 West Lombard St, Baltimore, Maryland, 21201, USA.
Tel: 001 410 706 8408
jian@cfs.umbi.umd.edu
cfs.umbi.umd.edu/cfs/people/JianZhang.htm

Specialty Keywords: Biosensing, Nanomaterial, Fluorescence.

My research interests focus on biosensing on nano-scale material including carbohydrate, DNA, and protein using fluorescence, especially surface enhanced fluorescence on metal particle. Emission can be enhanced more effectively on the coupled metal particles. These results are used on *in viva* target biomaterial detection and single molecule image in living cell.

- Zhang, J.; Gryczynski, I.; Gryczynski, Z.; Lakowicz, J. R. Dye-labeled silver nanoshell-bright particle Journal of Physical Chemistry B 2006 110, 8986-8991.
- Zhang, J.; Lakowicz, Joseph J. R. Model DNA detection by metal enhanced fluorescence from immobilized silver nanoparticles on solid substrate Journal of Physical Chemistry B 2006 110, 2387-2392.

Date submitted: 10th August 2006

Yongxia Zhang, Ph.D.

Institute of Fluorescence,
Laboratory for Advanced Fluorescence Spectroscopy,
Medical Biotechnology Center, N249,
University of Maryland Biotechnology Institute,
725 West Lombard St., Baltimore, Maryland 21201, USA.
Tel: 410 706 4566   Fax: 410 706 4600
zhangy@umbi.umd.edu

Specialty Keywords: Biosensors, Nanotechnology,
Phosphorescence.

My research interests focus on development and application of plasmonic/phosphorescence–based biosensors using noble metallic nanoparticles. Triplet states can be significantly enhanced on the metal nanopartilce surface. These finding can be used to promote triplet-based assays, such as those used in photodynamic therapy.
-Zhang, Y., Aslan K, Malyn N.S. and Geddes, C.D. "Metal-Enhanced Phosphorescence (MEP)" Chem. Phys. Lett. , 2006, 427, pp. 432-437.

-Zhang, Y., Johnson. D.T. "Comparison of different alcohols in a caged lumophore for measuring supersaturated dissolved oxygen" Analytica Chimica Acta (2004), 511(2), pp. 333-337.

Date submitted: 8th November 2006

Jie Zheng, Ph.D.

Department of Physiology and Membrane Biology,
University of California at Davis,
School of Medicine,
Davis, CA 95616, USA.
Tel: 530 752 1241   Fax: 530 752 2652
JZheng@UCDavis.edu

Specialty Keywords: FRET, Patch Fluorometry, Fluorescent
Proteins, Ion Channels, Signal Transduction.
AIM 2005 = 13.8

Using novel fluorescence techniques in combination with electrophysiology and molecular biology to better understand membrane protein structure and dynamic rearrangements in the structure that underlie cellular signal transduction. The focus of my research is ion channels that form the basis of electrical excitability of neurons and muscle cells. Fluorophores are attached to the moving parts of the channel as molecular sensors to detect structural changes in real time.
-Zheng, J. (2006) Patch fluorometry: shedding new light on ion channels, *Physiology*, 21, 6-12
-J. Zheng, and W.N. Zagotta (2003). Patch-clamp fluorometry recording of conformational rearrangements of ion channels. *Science's STKE*, pl7.
-J. Zheng, and W.N. Zagotta (2000). Gating rearrangements in cyclic nucleotide-gated channels revealed by patch-clamp fluorometry. *Neuron*, 28, 369-374.

Date submitted: 13th October 2006

Alexander Zilles, Ph.D.

ATTO-TEC,
Am Eichenhang 50,
Siegen, 57076,
Germany.
Tel: +0049 271 740 4735
zilles@atto-tec.de
www.atto-tec.com

Specialty Keywords: Fluorescence, Fluorescent Organic Dyes, Biolabeling, Bioanalytics, Photofading, UV-absorber.

My research interests are based on the design and synthesis of novel fluorescent dyes. Individual functionalisation of these dyes make them highly suitable for bioanalytical applications e.g. biolabeling of nucleotides, proteins, etc.

I am further interested in the development and investigation of detergent additives to prevent photodegradation of dyed fabrics in particular cellulosic based fibers e.g. cotton.

-Arden-Jacob J., Frantzeskos J., Kemnitzer N. U., Zilles A., Drexhage K.H., Spectrochim. Acta 57A, 2271-2283 (2001). Zilles A., PhD Theses "The design and synthesis of detergant additives for the photo-chemical protection of dyed fabrics". University of Leeds, Department of Colour Chemistry (2002).

Date submitted: 17th August 2005

Victor N. Zozulya, Ph.D.

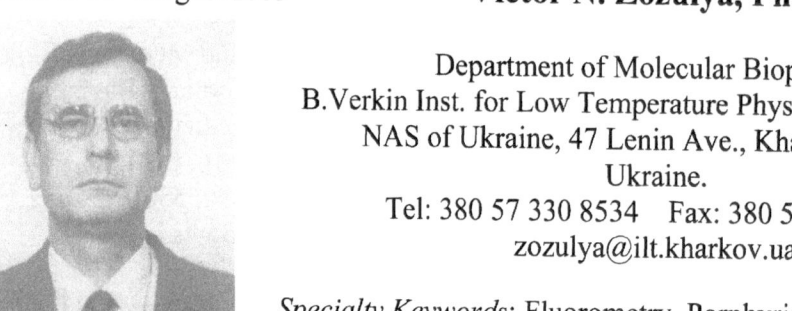

Department of Molecular Biophysics,
B.Verkin Inst. for Low Temperature Physics & Engineering,
NAS of Ukraine, 47 Lenin Ave., Kharkov, 61103,
Ukraine.
Tel: 380 57 330 8534   Fax: 380 57 345 0593
zozulya@ilt.kharkov.ua

Specialty Keywords: Fluorometry, Porphyrin-Nucleic Acid Complexes, Phenazine-porphyrin Conjugates.
AIM 2003 = 2.2

Investigation of fluorescent and binding properties of phenazine and porphyrin dyes in complexes with polynucleotides and nucleic acids. Utilization of phenazine-porphyrin covalent conjugates as fluorescent probes and stabilizers of telomeric G-quadruplexes.

-O.A. Ryazanova, V.N. Zozulya, I.M. Voloshin, V.A. Karachevtsev, V.L. Makitruk, S.G. Stepanian. *Spectrochim. Acta*, 60A, 2004, 2005-2011.
-V. Zozulya, Yu. Blagoi, I. Dubey, D. Fedoryak, V. Makitruk. O. Ryazanova, A. Shcherbakova. *Biopolymers (Biospectroscopy)*, 72 (4), 2003, 264-273.

Date submitted: 15th June 2005

Werner Zuschratter, Ph.D.

Special Laboratory Laserscanning Microscopy,
Leibniz Institute for Neurobiology,
Brenneckestr. 6, Magdeburg,
Saxony-Anhalt, 39118, Germany.
Tel: +49 391 626 3331   Fax: +49 391 626 3328
Zuschratter@ifn-magdeburg.de
www.ifn-magdeburg.de

Specialty Keywords: Multi-dimensional Microscopy, Cell Microscopy, TSCSPC, FLIM & FLIN, FRET.

W.Z. is head of the special laboratory (SL) for Electron- and Laserscanning- Microscopy at Leibniz Institute for Neurobiology since 1992. The SL provides high resolution microscope facilities and service for multi-dimensional microscopy with emphasis on fluorescence lifetime spectroscopy & microscopy and FRET analysis based on time- and space correlated single photon counting. Applications mainly focus on minmal invasive observations of protein-protein interactions within living neurons using non-scanning DL- /QA imaging detectors.

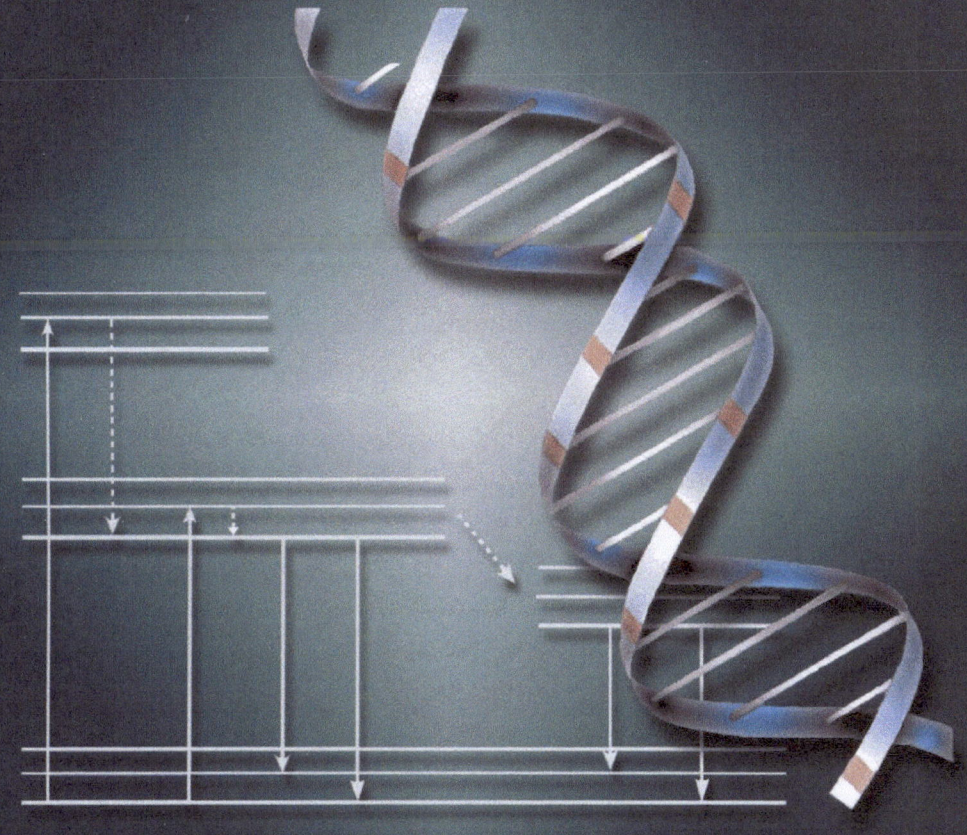

2007

Companies
In Fluorescence

Company & Institution Entries

BERTHOLD TECHNOLOGIES
GmbH. & Co. KG.

Bioanalytical Instruments,
P.O. Box 100 163,
75312 Bad Wildbad,
Germany.
Phone: +49 7081 177 - 0
Fax: +49 7081 177 - 100
Bio@Berthold.com
www.Berthold.com/Bio

Microplate Fluorometers, Imaging Instruments for Fluorescence Applications

BERTHOLD TECHNOLOGIES provides a variety of instruments for your fluorescence applications: The dedicated microplate fluorometer Twinkle LB 970, the multimode reader Mithras LB 940, the gel documentation system NightHAWK LB 984 and the imaging system NightOWL LB 981.

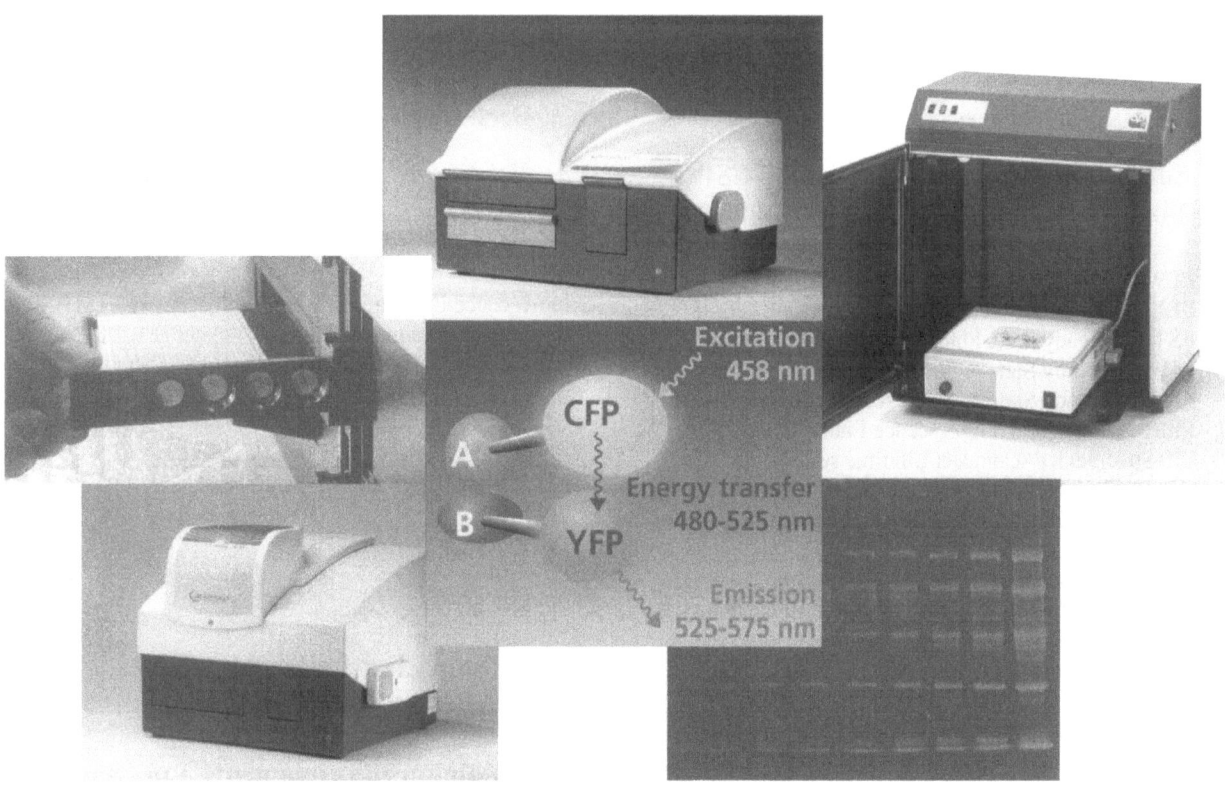

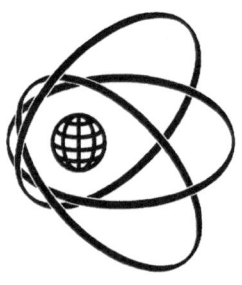

Boston Electronics Corporation

91 Boylston Street, Brookline, Massachusetts 02445, USA.

(800)347-5445 or (617)566-3821   fax (617)731-0935
www.boselec.com   tcspc@boselec.com

Boston Electronics are exclusive North American agents for the fluorescence lifetime measuring modules of Becker & Hickl GmbH of Berlin; for the fluorescence lifetime measuring spectrometers of Edinburgh Instruments Ltd. of Edinburgh; and for the Photon Counting APDs (SPADs) of id Quantique SA, Geneva.

Contact us for:

Time Correlated Single Photon Counting (TCSPC) Modules and Spectrometers, Lifetime and Steady State spectrofluorimeters, Picosecond Diode Lasers, MCPs and PMTs, Photon Counting APDs (SPADs), Gated Photon Counters and Multiscalers.

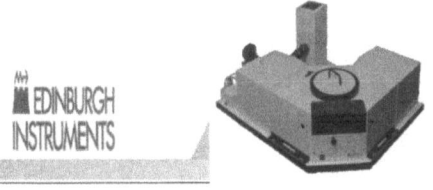

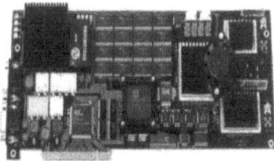

Becker & Hickl TCSPC Modules and Systems

Edinburgh Instruments Modular TCSPC Lifetime and Steady State Spectrometers

FL920   Fluorescence Lifetime
FLP920 Fluorescence & Phosphorescence Lifetime
FS920   Steady State Fluorescence
FSP920 Steady State & Phosphorescence Lifetime
FLS920 Steady State & Fluorescence Lifetime
FLSP920 Complete Fluorescence Laboratory
LifeSpec Series Compact Lifetime series
OB920 Series   Dedicated Lifetime systems
Mini-Tau Series Miniature Lifetime systems

Edinburgh Instruments Analytical Division is a world leader in the design and manufacture of single photon counting sensitive steady state and time-resolved fluorescence and phosphorescence spectrometers.

Laser Flash Photolysis spectrometers measure laser induced transient absorption and emission with temporal resolution from nanoseconds to seconds together with the associated emission spectra.

Becker & Hickl sets the standard, offering TCSPC electronic modules for applications including:

Fluorescence Lifetime Imaging Microscopy (**FLIM**)
Fluorescence Resonance Energy Transfer (**FRET**)
Fluorescence Correlation Spectroscopy (**FCS**)
Single Molecule Detection

TCSPC Modules	TCSPC Systems
SPC-130/4	**Simple-Tau**
SPC-140/4	**PML-Spec**
SPC-150/4	
SPC-630	
SPC-830	

Boston Electronics Corporation is the exclusive North American agent for Becker & Hickl GmbH. Edinburgh Instruments, Ltd. & id Quantique SA

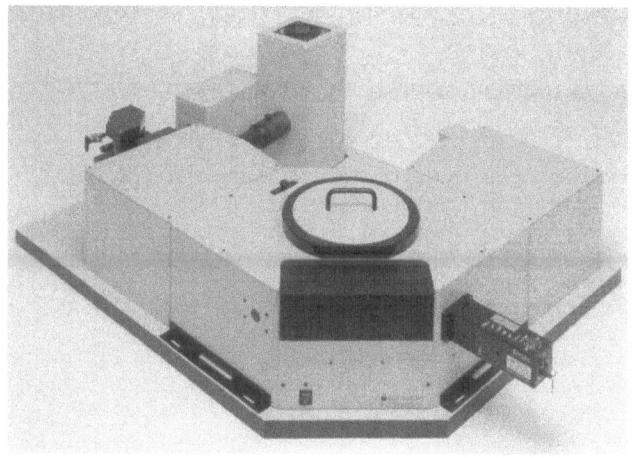

LAMBERT INSTRUMENTS
Turfweg 4
9313 TH Leutingewolde
The Netherlands.
Tel: +31 50 5018461 Fax: +31 50 5010034
lkvgeest@lambert-instruments.com
www.lambert-instruments.com

Specialty Keywords: Fluorescence Lifetime Imaging Microscopy - FLIM, Fluorescence Resonance Energy Transfer - FRET, frequency domain, LED

Lambert Instruments specializes in low light level image detectors and systems for scientific applications making use of image intensifiers, standard and custom made.

The **LIFA Fluorescence Lifetime Imaging Attachment** is a system that can be attached to any wide field fluorescence microscope, allowing fluorescence image acquisition and the generation of lifetime images.

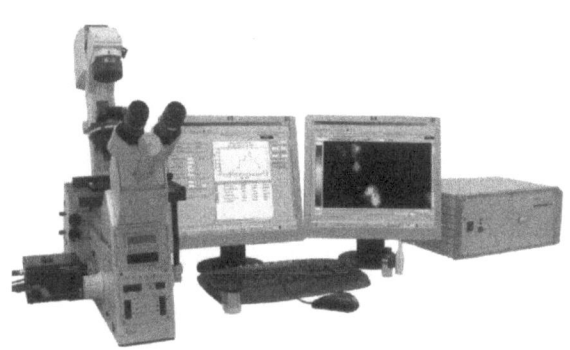

The LIFA system works in the frequency domain, giving a very efficient use of the available photons.

The use of LEDs as modulated light source makes the system reliable, easy to operate and very cost effective.

A high-resolution image intensifier that can be modulated up to 120 MHz is efficiently coupled to a digital CCD camera at the detection side.

The system is used among others in fundamental cell biology research and cancer research. With FLIM as the technique to detect fluorescence resonance energy transfer (FRET), macro molecular interactions inside cells can be easily detected.

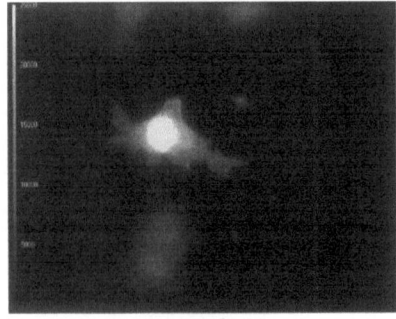

Intensity

Mammalian cells showing FRET at the plasmamembrane by a change of fluorescence lifetime.

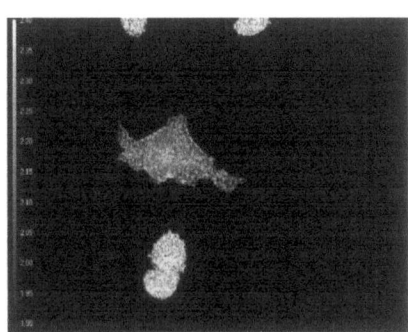

Lifetime

Ocean Optics, Inc.

Specialty Keywords

Fluorescence Spectrometers
Fluorometers
Spectrofluorometers
Time-gated Fluorescence
Excitation Sources
Linear Variable Filters
Sampling Accessories

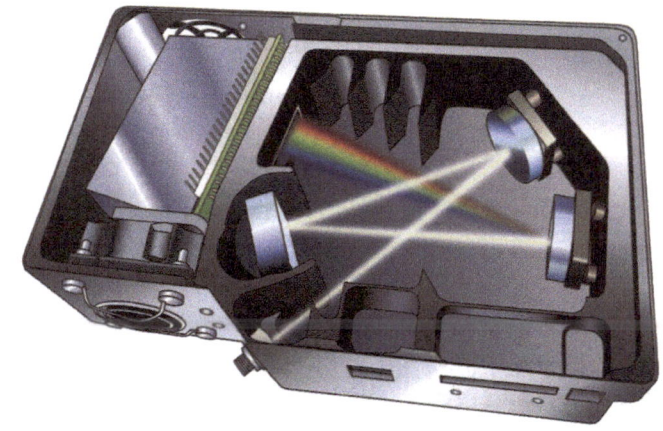

Company Description

Ocean Optics offers high-sensitivity spectrometers and optical-sensing accessories that combine with microscope couplers, X-Y stages, microtiter plates and other sampling fixtures to create systems for fluorescence and absorbance applications. Among our innovations is the QE65000 Spectrometer (optical bench pictured, above), a high-sensitivity spectrometer for low light-level applications. We also offer quantum-dot semiconductor nanocrystals for microbiology research. These quantum dots are an ideal tagging tool for biotechnologists, reagent suppliers and assay platform developers.

Contact Information

Ocean Optics, Inc.
830 Douglas Avenue
Dunedin, FL 34698

Phone: 727.733.2447
Fax: 727.733.3962
E-mail: Info@OceanOptics.com
OceanOptics.com

Ocean Optics, B.V.
Geograaf 24
6921 EW Duiven
The Netherlands

Phone: +31 (0) 26 319 0500
Fax: +31 (0) 26 319 0505
E-mail: Info@OceanOpticsBV.com
OceanOpticsBV.com

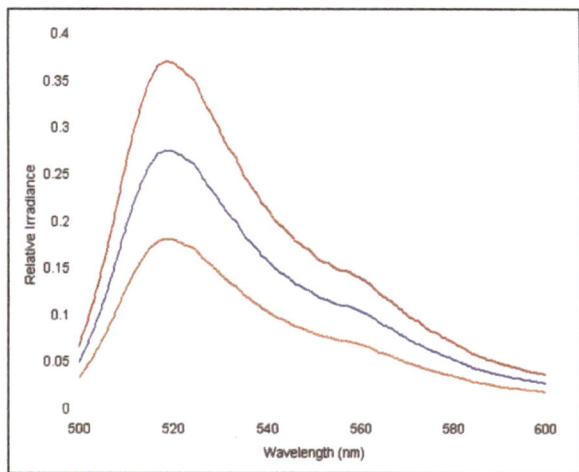

PCO AG
Donaupark 11
93309 Kelheim
Germany
fon: +49 (0)9441 2005 0
fax: +49 (0)9441 2005 20
info@pco.de, www.pco.de
in Amerika: www.cookecorp.com

Speciality Keywords: cooled CCD camera, sensitive camera, scientifc imaging, low light camera, emCCD camera, low noise camera

Company

In 1987, PCO AG was founded with the objective to develop and to produce specialized fast and sensitive video camera systems, mainly for scientific applications. Meanwhile the product range of PCO cameras covers digital camera systems with high dynamic range, high resolution, high speed and low noise, which are sold in the scientific and industrial market all over the world.

Currently PCO is one of the leading manu-facturers of scientific cameras. Worldwide representatives, together with our own sales department and technical support assure that PCO keeps in touch with our customers and their needs. The actual wide range of specialized camera systems is the result of technical challenge and product specific know-how.

Products

The product range is structured into the following sections:
■ sensitive cameras
  12bit & 14bit dynamic CCD camera systems with high sensitivity and low noise
■ intensified cameras
  high resolution MCP image intensifier camera systems
■ speed cameras
  ultra- & high speed CMOS and image intensifier camera systems
■ specialized cameras
  CCD camera systems with double shutter features or modulation capabilities

Furthermore PCO develops customer specific solutions (OEM), like camera systems for industrial applications, where high dynamics and image quality are important issues.

Committment to excellence

A technical design according to advanced technolgies, a high standard of production, and strict quality controls guaranty a reliable performance of our cameras. Our own developments in conjunction with an excellent contact to leading manufacturers of image sensors ensure our access to state-of-the-art CCD-and CMOS-technology for our cameras.

PicoQuant GmbH

Rudower Chaussee 29,
12489 Berlin,
Germany.
Tel / Fax: +49 (0)30 6392 6560 / 6561
info@picoquant.com
www.picoquant.com

Special Keywords: **Pulsed Diode Laser, Photon Counting Instrumentation, Fluorescence Lifetime Spectrometer, Confocal Fluorescence Microscopes**

PicoQuant GmbH is a research and development company based in Berlin-Adlershof, Germany. The company is leading in the field of single photon counting applications. The product line includes:

Picosecond Pulsed Diode Laser Systems
- 370 to 1550 nm, repetition rate up to 80 MHz, pulse widths down to 50 ps
- High power laser at 780, 980, 1060 nm, with SHG for 490 and 530 nm
- Sub-ns pulsed LEDs from 265 to 600 nm
- Single or (computer controlled) multi-channel versions

Fluorescence Lifetime Spectrometer
- Picosecond to millisecond time resolution
- Modular and flexible design
- Advanced data analysis software

Confocal Fluorescence Microscopes
- Single molecucle sensitivity, 3-D scanning capability
- FLIM & FCS upgrade for Laser Scanning Microscopes
- Comprehensive analysis software

PC modules for TCSPC / Multi-channel Scaling
- Picosecond to millisecond timing
- Time-tagged mode & FCS option
- 4 channel routing capability

Services:1) Annual International Workshop on Single Molecule Detection/Spectroscopy and Ultrasensitive Analysis in Life Sciences and 2) European Short Course on Principles and Applications of Time-resolved Fluorescence Spectroscopy.

FREQUENCY GENERATORS, AGILE, QUIET, FAST

FREQUENCY SYNTHESIZERS

Accurate, stable frequencies on command, µs switching. For NMR, Surveillance,
ATE, Laser, Fluorescence. Low noise/jitter.   Adapting to your needs with options.
Now to 6.4 GHz Clean, Fast Switching

FREQUENCY SYNTHESIZERS

Model	Frequency Range	Resolution	Switching Time	Phase-Continuous Switching	Rack-Mount Cabinet Dimensions[1]	Remote-Control Interface	Price Example[2]
PTS 040	0.1-40 MHz	optional 0.1 Hz to 100 KHz	1-20µs	optional	5.25" H x 19" W	BCD(standard) or GPIB(optional)	$5,330.00
PTS 120	90-120 MHz	optional 0.1 Hz to 100 KHz	1-20µs	optional	5.25" H x 19" W	BCD(standard) or GPIB(optional)	$5,330.00
PTS 160	0.1-160 MHz	optional 0.1 Hz to 100 KHz	1-20µs	optional	5.25" H x 19" W	BCD(standard) or GPIB(optional)	$6,495.00
PTS 250	1-250 MHz	optional 0.1 Hz to 100 KHz	1-20µs	optional	5.25" H x 19" W	BCD(standard) or GPIB(optional)	$7,440.00
Type 1 PTS 310 Type 2	0.1-310 MHz	1 Hz	1-20µs	standard	3.50" H x 19" W	BCD(standard) or GPIB(optional)	$6,425.00 $5,850.00
PTS 500	1-500 MHz	optional 0.1 Hz to 100 KHz	1-20µs	optional	5.25" H x 19" W	BCD(standard) or GPIB(optional)	$8,720.00
PTS 620	1-620 MHz	optional 0.1 Hz to 100 KHz	1-20µs	optional	5.25" H x 19" W	BCD(standard) or GPIB(optional)	$9,625.00
PTS 1600	1-1600 MHz	1 Hz	1-20µs	optional	5.25" H x 19" W	BCD(standard) or GPIB(optional)	$10,550.00
PTS 3200	1-3200 MHz	1 Hz	1-20µs	standard	5.25" H x 19" W	BCD(standard) or GPIB(optional)	$13,350.00
PTS 6400	1-6400 MHz	1 Hz	1-20µs	standard	5.25" H x 19" W	BCD(standard) or GPIB(optional)	$15,550.00
PTS x10	user specified 10 MHz decade	1 Hz	1-5µs	standard	3.50" H x 19" W	BCD(standard) or GPIB(optional)	$3,000.00
PTS D310	two channels 0.1-310 MHz	0.1 Hz	1-20µs	standard	5.25" H x 19" W	BCD(standard) or GPIB(optional)	$8,560.00
PTS D620	two channels 1-620 MHz	0.1 Hz/0.2 Hz	1-20µs	standard	5.25" H x 19" W	BCD(standard) or GPIB(optional)	$13,240.00

[1] Bench cabinets are 17" wide

[2] Prices are U.S. only and include manual and remote (BCD) control,
1 Hz resolution where optional, OCXO frequency standard
(PTS 1600, 3200 and PTS 6400 digital front panel)

**PTS CAN SUPPLY OEM-TYPE SYNTHESIZERS FOR ALL
FLUORESCENCE AND OTHER SCIENTIFIC APPLICATIONS**

PROGRAMMED TEST SOURCES, INC.

P.O. BOX 517, 9 Beaver Brook Road
Littleton, Massachusettts 01460
Telephone: 978 486 3400
Fax: 978 486 4495
http://www.programmedtest.com
email: sales@programmedtest.com

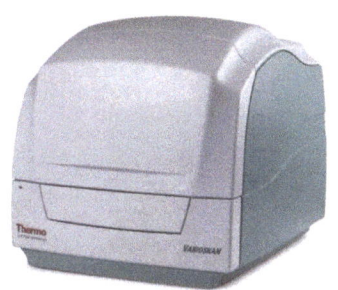

Thermo

ELECTRON CORPORATION
Microplate Instrumentation
Ratastie 2
01620 Vantaa
Finland
tel: +358-9-329-100
fax: +358-9-32910-415
www.thermo.com/mpi

Specialty keywords: **Spectral scanning, Spectrofluorometer, Spectrophotometer, Fluorescence, Luminescence, Absorbance, FRET, BRET, TRF, Fluorescence intensity, Fluorescence Polarization, Multimode reader, Microplate reader, Plate reader, Microplate.**

Thermo Electron brings over 20 year's of experience in fluorometric microplate reading to you with

- Varioskan® Flash - spectral scanning multimode reader
- Appliskan™ - filter based multimode reader
- Fluoroskan Ascent® F L- combined microplate fluorometer/luminometer
- Fluoroskan Ascent® - dedicated microplate fluorometer
- Luminoskan Ascent® - dedicated microplate luminometer

Covering a wide range of **applications from cell proliferation, reporter gene assays to DNA quantification and immunoassays,** Thermo has a powerful reader solution for you. Thermo's readers offer you:

- **Excellent performance.** High quality optical systems guarantee highly sensitive and reliable readings.
- **In-built dispensers**. For optimization of kinetic reactions.
- **Advanced but easy to use PC software**. For flexible instrument control and powerful data management.
- **Easy robotics integration.** For increased throughput.

For more information about Thermo's microplate instrumentation offering, visit **www.thermo.com/mpi.**

About Thermo

Thermo Electron Corporation is the world leader in analytical instruments. Our instrument solutions enable our customers to make the world a healthier, cleaner and safer place. Thermo's Life and Laboratory Sciences segment provides analytical instruments, scientific equipment, services and software solutions for life science, drug discovery, clinical, environmental and industrial laboratories. Thermo's Measurement and Control segment is dedicated to providing analytical instruments used in a variety of manufacturing processes and in-the-field applications, including those associated with safety and homeland security. For more information, visit www.thermo.com.

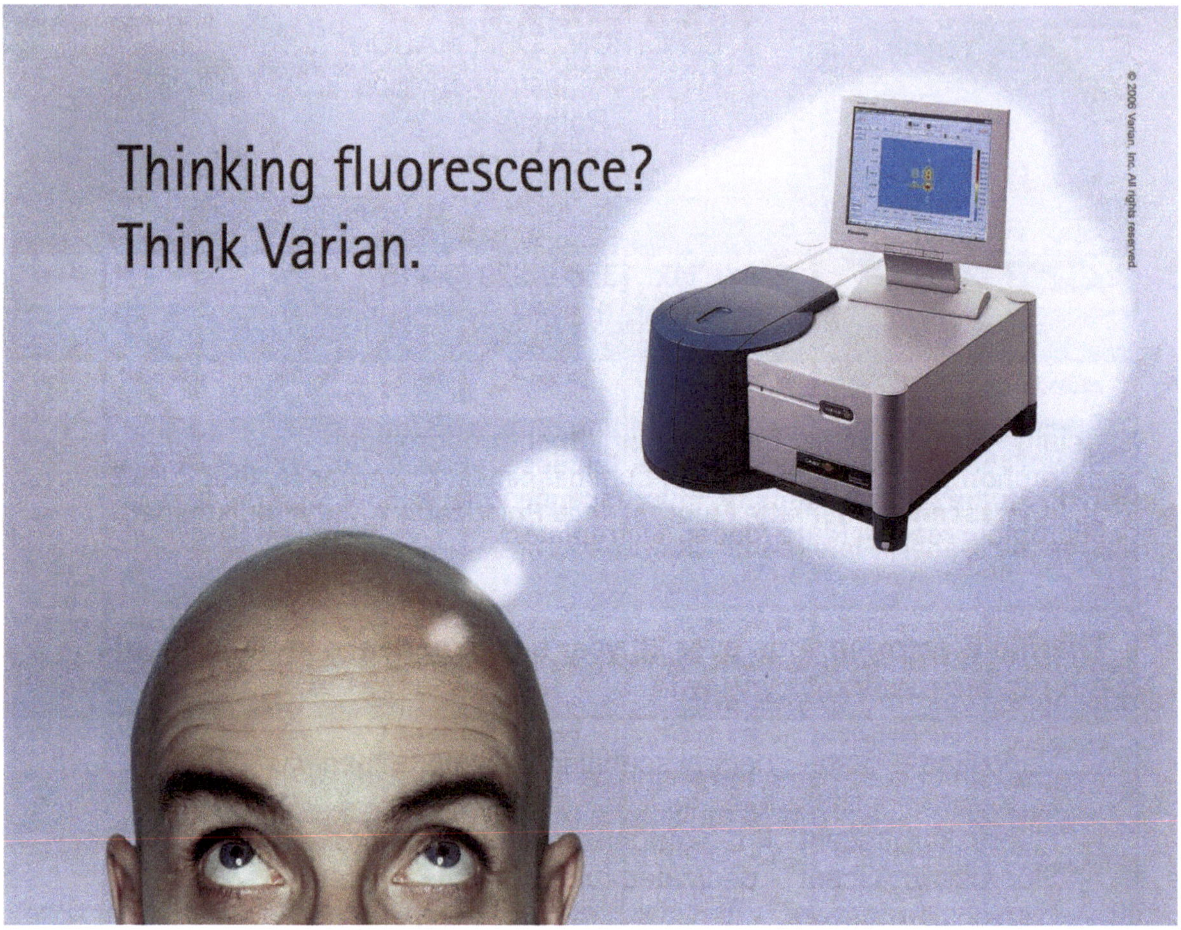

Scientists and workers in academia, industry or government employing fluorescence in their everyday working lives are invited to apply for entry in the *Who's Who in Fluorescence* 2008 annual volume.

The annual volume, edited by Dr's Chris D. Geddes and Joseph R. Lakowicz, publishes the names, addresses, contact details and a brief paragraph describing fluorescence workers specialities.

To apply for entry in the Who's Who in Fluorescence 2008 volume, complete the personal template (Word format) obtained on request from the Who's Who in Fluorescence Annual Volume Co-ordinator: Caroleann Aitken at wwif@umbi.umd.edu and returned to the same, no later than August 31st 2007. Unsuccessful entries, entries not conforming to the template format, or those received after the closing date will be returned without further consideration.

Contributors are asked to keep file sizes as small as possible by using appropriate standard picture formats, such as JPEG and TIFF etc. Alternatively, electronic versions can be submitted by mail (CD) to:

Dr. Chris D. Geddes
Founding Editor-in-Chief: *The Who's Who in Fluorescence,*
The Institute of Fluorescence,
Medical Biotechnology Center, N249,
725 West Lombard Street,
Baltimore, Maryland, 21201, USA.

Galley proofs will no longer be posted on the Who's Who in Fluorescence website as in previous years. Subsequently, we ask authors to be extra vigilant in the preparation of their entries.

Personal half-page entries in the Who's Who in Fluorescence 2008 volume are free of charge.

Fluorescence-based companies may also submit a full-page company profile in the Who's Who in Fluorescence 2008 volume for a fee of $600.00 (black and white), $2000.00 (4-colour), prices subject to change. Full-page company templates can be obtained on request from the Who's Who in Fluorescence Annual Volume Co-ordinator: Caroleann Aitken at wwif@umbi.umd.edu. For colour images and high resolution images, companies are asked to contact the editors to discuss their requirements beforehand.

Institutions, academic research groups and centres of scientific excellence are also invited to submit full-page profiles for a fee of $250.00 (black and white), $2000.00 (4-colour), also using the company template. Both company and institutional submissions are also to be submitted by August 31st 2007.

Further enquiries are to be directed to the editor or WWIF Co-ordinator, Caroleann Aitken, at wwif@umbi.umd.edu

Author Impact Measure (AIM): An author publication statistic for the *Who's Who in Fluorescence Volume*

The *Who's who in Fluorescence* Annual Volume employs a voluntary personal publication statistic, which first appeared in the 2005 volume.

Contributors are asked to calculate their *Author Impact Measure* (AIM) number and supply this along with their completed Who's Who in Fluorescence template via the usual e-mail address: wwif@umbi.umd.edu or Caroleann@cfs.umbi.umd.edu no later than August 2007, for the 2008 volume.

The AIM number is expected to show both an author's progress and productivity in a given year. This statistic will be published in the Who's Who in Fluorescence volume, on a *voluntary* basis. The AIM number is simply calculated as the cumulative *impact number* (from the ISI database) of Journals published in, in a single year, multiplied by the frequency of publications.

For example, if an contributor published 3 papers in the *Journal of Fluorescence* and 1 paper in the *Journal of Physical Chemistry B* in the year 2002. The 2002 AIM number would be:

$$3*(0.761) + 1*(3.611) = \underline{5.894}$$

Aim numbers are to be calculated only for articles published at the time of submission (not pending or in press). AIM numbers for up to 3 years previous to the Who's Who in Fluorescence Annual publication date can be quoted, i.e. the 2008 volume will publish one AIM number, selected from any *one year* between 2005 and 2007 inclusive, at the contributor's discretion.

We hope you find this author publication statistic informative and we look forward to any suggestions you may have.

Kind Regards,

Dr Chris D. Geddes,
Professor,
Founding Editor-in-chief: *The Who's Who in Fluorescence Annual Volume.*

Personal Template

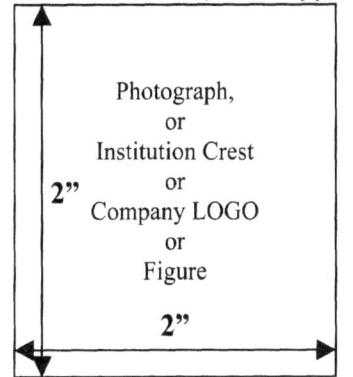

Company and Institutional Template

Company Name / Institution (14 pt Times New Roman, **Bold**)

Department, Institution, Branch
Street name, City,
County, Zip code,
Country.
Tel: Fax Numbers
E-mail Address
Homepage URL

Specialty Keywords: **Keyword 1, Keyword 2, Keyword 3**

Letter page size, 8.5x11 in (Portrait), 1" left, 1" right margins. The total area *should not exceed* 8.5 *in* height x 6.5 *in* width. Where possible text should be 12 pt Times New Roman.

Please submit entries as a word file, if PDF file is used please ensure the file is not locked, to allow for type setting.

THIS SPACE MAY BE FULLED AS REQUIRED.
(Lay out, text and graphics to the companies requirements)

4 Color entry = $2000.00 usd
Black and White entry = $600.00 usd

For further information please contact the Who's Who in Fluorescence Annual Volume Co-ordinator: *Caroleann Aitken*,

Assistant to the Founding Editor-in-Chief: Professor Chris D. Geddes.

e-mail: WWIF@UMBI.UMD.EDU

Company and institutional entries will appear at the back of the issue, also alphabetically. Companies and institutions occupy one *Journal* Page respectively.

E-mail complete forms by early August 2007 to:
wwif@umbi.umd.edu

8.5 "

6.5 "